Limit States Design of Structural Steelwork

Limit States Design of Structural Steelwork

Third edition

David A. Nethercot

Professor of Civil Engineering and Head of
Department of Civil & Environmental
Engineering
Imperial College of Science, Technology &
Medicine, UK

London and New York

First published 1986 by Van Nostrand Reinhold (UK) Co. Ltd.
Second edition published 1991 by Chapman & Hall
Reprinted 1994, 1995, 1996

Third edition published 2001 by Spon Press
11 New Fetter Lane, London EC4P 4EE
Simultaneously published in the USA and Canada by Spon Press
29 West 35th Street, New York, NY 10001

Spon Press is an imprint of the Taylor & Francis Group

© 1986, 1991, 2001 D. A. Nethercot

Typeset in 10/12pt Times by Wearset, Boldon, Tyne and Wear
Printed and bound in Great Britain by Biddles Ltd, Guildford and
King's Lynn

British Library Cataloguing in Publication Data
A catalogue for this book is available from the British Library.

Library of Congress Cataloging-in-Publication Data
Nethercot, D. A.
    Limit states design of structural steelwork/David A. Nethercot.
    – 3rd ed.
        p. cm.
    ISBN 0-419-26080-3 (alk. paper) – ISBN 0-419-26090-0 (pbk. :
alk. paper)
    1. Steel, Structural. 2. Plastic analysis (Engineering) I. Title.

TA684 .N43 2001
624.19821–dc21   00-055646

ISBN 0-419-26090-0

10 0623253九

Contents

Preface to the third edition

Although the package of structural Eurocodes will eventually replace national documents, such as BS 5950, that situation will not be reached for some years. It was therefore decided in 1997 by the BSI Committee responsible for the design aspects of BS 5950 that a technical amendment should be produced. This was justified largely on the basis of safety. Improved technical understanding gained through the consultative process used to prepare Eurocode 3 had identified certain areas of the code where technical change was desirable; further experience in use had revealed a number of instances where some re-wording could improve clarity and avoid inconsistent interpretations; it was also the case that several of the supporting product standards had been changed, with the result that cross-referencing needed correction.

This third edition of 'Limit States Design of Structural Steelwork' takes full account of all the changes to the Part 1 of BS 5950 covered by the 2000 Amendment. Arrangements have recently been put in place to prepare amendments to the Part 3.1 that deals with Composite Construction and the Part 8 that covers Fire Resistant Design. Thus, when these documents are completed, Chapters 9 and 12 may well need some further revision. Eventually, when EC3 supersedes BS 5950 as the main reference code for steel construction for the UK (and virtually the whole of Europe), further revisions will be necessary. For the next few years, however, the contents of this volume should be regarded as being fully in accordance with both the principles and the practice of structural steel design as it operates in the UK.

In preparing the manuscript, I received valuable assistance from both my secretary at the University of Nottingham, Sue Muggridge and, since moving to Imperial College in 1999, my current secretary, Valerie Crawford. I am also grateful to the members of the relevant BSI Committee, B/525/31, of which I am now Chairman, as well as my many contacts, in both the UK and overseas, in the structural steelwork community, who have, knowingly or unknowingly through discussion, correspondence and their own writings, contributed to my understanding of the subject area on which this book is based.

D. A. Nethercot

Preface to the second edition

The publication of the first set of amendments to BS 5950 (in the form of the 1990 reprinting of the Code), together with the appearance of Part 3.1 and Part 8, has provided the motivation for revising this text. Thus in addition to a general updating of the previous version, the second edition contains new chapters to introduce the principles of 'composite construction' (Chapter 9) and to explain 'design for fire resistance' (Chapter 12). These are based on the treatment of these subjects in Parts 3.1 and 8 respectively. An opportunity has also been taken to revise and expand the original material on joints and frames with the result that Chapters 7 and 8 and Chapters 10 and 11 now provide a significantly enhanced coverage of these topics.

Since completion of the first edition I have been drawn more closely into the BSI network of committees dealing with both BS 5950 and the UK input to the forthcoming Eurocodes. I was appointed to the main CSB/27, responsible for all parts of BS 5950 in 1986, and am currently that committee's only academic member. Publication of the Part I in September 1985 saw the start of a series of courses and workshops organized by the Steel Construction Institute to explain the background to the new code. I have lectured on these on more than 50 occasions – sometimes outside the UK. The experience provided by these BSI and SCI activities has been invaluable in preparing this second edition.

During the summer of 1990 I was fortunate to spend some time as a visiting professor in the Institut pour Construction Metallique at the Ecole Polytechnique Federale de Lausanne. The Swiss scenery and the early morning start in Professor J-C Badoux's institute supplied the combination of creative environment and industry within which much of the work on this new edition was conducted. However, the manuscript was actually prepared in the University of Nottingham and particular thanks are therefore due to my secretary, Sue Muggridge.

D. A. Nethercot

Preface to the first edition

The tittle 'BS 449' is recognized throughout the world as the main British code of practice devoted to the design of structural steelwork. First published as a byproduct of the activities of the Steel Structures Research Committee in 1931, BS 449: *The Structural Use of Steel in Buildings* has been revised, extended and amended to take account of improved understanding of structural behaviour, changes in fabrication techniques and the requirements of new forms of construction on several occasions. The most recent metric edition is dated 1969. Some two years prior to this a decision was taken to begin work on a completely revised version which would not only update the document's detailed design procedures but would recast these into the more progressive limit states format. Of course such a move was not universally well received by structural designers; it is still unpopular with a section of the profession today. It did, however, represent a course of action that either has been or is being pursued by most of the main UK structural codes as well as by steelwork codes in many other parts of the world.

The author first became directly involved in the production of this new code in 1971. It was through this contact that the idea for a textbook explaining the material of the code to both students and practising engineers gradually developed. Work on the text began at about the time that the original draft for comment – the so-called B/20 Draft – was issued in 1977. Because the reaction to B/20 was sufficient to require substantial alterations and re-drafting, the appearance of the code with its new designation BS 5950 has taken several years. Completion of the text has, of course, had to await finalization of this document.

The book is not intended to be a commentary upon the new code; that document has been prepared by Constrado as part of their role in producing the actual text for both the code and the supporting material. Rather, it is a textbook on structural steel design according to the principles and procedures of BS 5950. As such it is aimed principally at students of steelwork design – whether they be undertaking courses in universities or polytechnics or, having successfully completed this phase of their career, are working in practice and want to update their knowledge. Therefore it is hoped that the material will be both self-contained and suitable for private study. It does, of course, make frequent reference to particular clauses in the code itself.

In writing this book the author has benefited enormously from various forms of interaction with a large number of organizations and individuals. These have included those responsible for the preparation and drafting of the code, teachers of steelwork design, representatives of overseas steelwork code committees, engineers in practice and delegates on various post-experience courses and seminars on either the new code or on steelwork design in general. Frequently, seemingly small points raised in discussion have provided an impetus for a change in the text or for the inclusion of an additional point of explanation. To all of these the author is most sincerely grateful.

The manuscript was prepared using facilities of the Department of Civil and Structural Engineering in the University of Sheffield. The text was typed by Miss Janet Stacey whose patience in dealing with the numerous revisions is greatly appreciated.

D. A. Nethercot

Notation

A	cross-sectional area
A_e	effective area of a tension member
A_g	gross area of section
A_n	net area of section
A_s	shear area of bolt
A_{sc}	steel area in compression
A_t	tensile area of a bolt
a	throat size of fillet weld, projections of baseplate, depth of haunch, distance between member axis and restraint axis, spacing of vertical web stiffeners
a_1	net area of connected leg of section, distance to topmost bolt hole
a_2	gross area of unconnected leg of section, distance to lowest bolt hole
B	width of section
B_e	effective breadth of concrete compression flange
b	clear width of plate
b_e	effective width of slender plate
b_1	length of stiff bearing
D	overall depth of section, hole diameter
D_p	depth of metal deck
D_s	depth of slab
d	diameter of bolt, clear web depth
d_v	spread of bolt holes
E	Young's modulus
E_{st}	strain hardening modulus
e	distance between centroid and extreme fibre of a section, end distance
e_x, e_y	eccentricity of axial load

F	axial load
F_q	difference between actual shear in web adjacent to stiffener and shear capacity of web
F_s	applied shear
F_t	applied tension
f	factor to allow for bending effects
f_a	axial stress
f_b	bending stress
f_{cu}	concrete strength
g	gauge of holes
h	storey height, depth of shear stud
h_{sc}	lever arm
h_1	height to eaves for a portal frame
h_2	height from eaves to apex for a portal frame
H_p	heated perimeter in metres
H_p/A	section factor
I	second moment of area
I_g	second moment of area of uncracked section
I_p	second moment of area of cracked section
I_s	second moment of area of web stiffener
I_y	second moment of area about y-axis
I_x	second moment of area about x-axis
K_e	factor used to define A_e, see equation (3.1)
K_s	factor on slip resistance of HSFG bolt, see equation (7.5)
k	effective length factor
k_{bs}	factor to allow for hole type
k_1, K_2	restraint parameters for column in a continuous frame
L	length
L_y	maximum distance between torsional restraints in a plastically designed structure, see equation (11.5)
L_u	maximum unbraced length for a member in a plastically designed structure
L_1, L_2	parts of block shear failure line
L_3	limiting spacing of compression flange restraints in a plastically designed frame
l	effective column length
M_{ax}	buckling resistance moment for combined axial load F and major-axis moment M_x
M_{ay}	buckling resistance moment for combined axial load F and major-axis moment M_y

M_b	buckling resistance moment
M_c	moment capacity of section
M_E	elastic critical moment for lateral-torsional buckling
M_p	fully plastic moment of cross-section
M_{rx}	reduced moment capacity about x-axis in the presence of axial load F
M_{ry}	reduced moment capacity about y-axis in the presence of axial load F
M_s	plastic moment capacity of steel section
m	equivalent uniform moment factor
m_{LT}	equivalent uniform moment factor for lateral-torsional buckling of beams
N	number of shear connectors
N_p	value of N required for full interaction
n	ratio F/AP_y
n_1	length under point load due to load dispersion, see equation (5.13)
P	axial load
P_c	compressive capacity
P_{cr}	elastic critical load
P_{cx}	axial capacity for xx buckling
P_{cy}	axial capacity for yy buckling
P_o	maximum shank tension for bolt
P_s	shear capacity of a bolt
P_{sb}	slip resistance of an HSFG bolt
P_t	tensile capacity of a bolt
P_v	shear capacity
P_w	web buckling capacity
P_y	squash load of column
P_z	capacity of vertical web stiffeners
p_a	axial stress
p_b	bending strength
p_{bs}	bearing strength of connected parts
p_{bc}	bearing strength of bolt in plate
p_{bx}	maximum bending stress due to M_x
p_{by}	maximum bending stress due to M_y
p_c	compressive strength
p_s	shear strength of bolt material
p_w	design strength of weld
p_y	design strength of steel
Q	load effects
Q_k	connector strength

Q_d	design strength of steel shear connector
q_c	elastic critical stress for shear buckling
q_w	shear buckling strength of the web
R	structural strengths (resistances)
R_s	Ap_y, resistance of steel beam
R_c	$0.45f_{cu}B_cD_s$, resistance to concrete flange
R_f	BTp_y, resistance of steel flange
R_w	R_s-$2R_f$, resistance of overall web depth
R_v	dtp_y, resistance of clear web depth
r	$(y_c$-$y_t)/d$, measure of web depth in compression, NQ_k/R_s, degree of shear connection
r_{aa}	radius of gyration of an angle about an axis through the centroid parallel to the gusset
r_{min}	minimum radius of gyration
r_{vv}	radius of gyration about minor principal axis
r_x	radius of gyration about x-axis
r_y	radius of gyration about y-axis
S_{rx}	reduced plastic section modulus about x-axis in the presence of axial load F
S_{ry}	reduced plastic section modulus about y-axis in the presence of axial load F
S_x	plastic section modulus about x-axis
s	leg length of fillet weld
s_p	staggered pitch of holes
T	flange thickness
t	plate thickness, web thickness
U_s	specified minimum ultimate tensile strength of steel
u	buckling parameter
V_b	shear buckling resistance of a web panel
V_f	flange dependent shear buckling resistance
V_w	simple shear buckling capacity of the web
v	lateral deflection, slenderness factor, shear per unit length
W	transverse load
W_p	value of W at plastic collapse
W_y	value of W at initial yield
w	pressure on underside of baseplate, distributed load on beam

x	torsional index
Y_s	specified minimum yield strength of steel
y	neutral axis depth
Z	elastic section modulus
z_1, z_2	index, see equation (6.4)
α	$2y_c/d$, measure of bending present in a compressed plate, see Table 8.1
α_s, α_L	modular ratio
β	M_1/M_2, ratio of end moments ($-1 \leq \beta \leq 1$)
$\beta_1 \beta_2 \beta_3$	limits on plate slenderness for plastic, compact and semi-compact cross-sectional behaviour respectively
γ_e	factor of safety in permissible stress design
γ_f	partial factor on loading
γ_m	partial factor on materials
γ_p	global load factor in plastic design, partial factor on structural performance
δ	deflection
ϵ	strain
ϵ_{sh}	strain at onset of strain hardening
ϵ_y	strain at initial yield
λ	slenderness
λ_c	l/r_{min} for main member of compound strut
λ_{cr}	load factor for elastic instability of frame
λ_{LO}	value of λ_{LT} below which $M_b = M_p$
λ_{LT}	$\sqrt{\dfrac{\pi^2 E}{p_y}} \sqrt{\dfrac{M_p}{M_E}}$, lateral-torsional slenderness
λ_w	$(0.6p_y/q_e)^{\frac{1}{2}}$, web slenderness
μ	slip factor
σ_p	stress at limit of proportionality
σ_{ult}	ultimate tensile stress
σ_y	yield stress
σ_{yL}	lower yield stress
σ_{yu}	upper yield point

Index of references to clauses in BS 5950

PART 1

Chapter 1

Steel as a structural material

Steel is the most widely used structural metal. Its popularity may be attributed to the combined effects of several factors, the most important of which are: it possesses great strength, it exhibits good ductility, it has high stiffness, fabrication is easy and it is relatively cheap. Good examples of structural steelwork design seek to exploit each of these features to the full [1].

Steel's high strength permits heavy loads to be carried by relatively small members, thereby reducing the self-weight of the structures. This reduction in dead load facilitates the construction of the large clear spans needed, for example in sports halls. At points of very high stress such as in the immediate vicinity of a bolt, yielding of the material will enable the load to be redistributed smoothly and safely; this process makes use of the property known as ductility. All structures will deform to a certain extent when loaded – even when such loading consists only of the structure's own self-weight. Because steel possesses great stiffness (as measured by its modulus of elasticity E) these deflections will not normally be large enough to require special consideration. Steel may be worked in the fabricating shop in a number of ways, for example sawing, drilling and flame cutting; it may also be joined together by welding. Finally the price of steel is substantially less than that of any possible competing metal; for instance aluminium costs about three times as much as the basic structural grades of steel.

In the civil engineering field steel is in competition principally with reinforced and prestressed concrete, timber and brickwork, with many designers seeing the usual choice as being simply between steelwork and concrete. The reasons most often given for selecting a steel structure are listed in Table 1.1. Since the requirements of individual projects may vary so much it is not possible to provide simple rules for selecting the 'best' solution, even on the limited basis of initial cost. Rather, the designer must consider each of the factors present, must decide on their relative importance and must then use his judgement and experience to decide upon the most appropriate solution.

As shown in Table 1.1, steelwork will often be the choice when questions of ease and speed of erection, for example a bridge over a busy railway line that can only be obstructed for short periods, large clear spans such as a grandstand for which no interference with visibility can be tolerated, or subsequent modifications to say a workshop which may be extended in size or into which additional cranage may be installed, are important. However, other factors, namely the non-availability of suitable aggregate locally or special architectural features, may also control the decision.

1.1 Production

Steel is made by refining iron which has itself been smelted from the basic iron ore in the blastfurnace. Ironmaking has changed little in principle in over 2000 years, although the actual techniques employed as well as the scale of production have, of course, altered considerably. Nowadays blast-furnaces operate continuously over a period of several years, producing up to 8000 tonnes of molten iron every 24 hours [2]. Iron ore, coke, limestone and sinter (a mixture of ore, coke and limestone that has previously been roasted together to remove some of the volatile matter) are fed into the top of the furnace. Air is blown through to increase the temperature, the oxygen content reacting with the hot carbon in the coke to form carbon monoxide which in turn releases the iron. The molten metal is periodically tapped off from near the base for subsequent use as the basic raw material employed in steelmaking.

1.1.1 Steelmaking

Four main processes exist for the production of steel. The oldest of these is the open-hearth process. Because it is slow and therefore relatively uneconomic it has largely been replaced by the basic oxygen (BOS)

Table 1.1 Advantages of steel structures

Item	Comments
Ease of erection	No formwork Minimum cranage
Speed of erection	Much of the structure can be prefabricated away from the site Largely self-supporting during erection
Modifications at a later date	Extensions/strengthening relatively straightforward
Low self-weight	Permits large clear spans
Good dimensional control	Prefabrication in the shop ensures accurate work

process [2] and the electric arc method [2], which was originally devised to produce high-quality steels requiring precise control over their composition. Today most structural steel is made using the BOS process shown in Figure 1.1. This commences with the operation known as charging, in which a mixture of molten iron and up to 30% scrap is poured into the top of the BOS vessel. High-purity oxygen is then blown in at great speed using a water-cooled lance. This combines with excess carbon and other unwanted impurities which then float off as slag.

During this time the temperature and chemical composition are carefully monitored and when both are adjudged correct the steel is tapped onto a ladle. At this stage a sample is taken for chemical analysis and subsequent examination of its physical properties; the results of these appear on the mill certificate which must be provided to the eventual purchaser of the steel. From the ladle the still molten metal is cast into moulds where it solidifies into the ingots which will be taken to the mill to be rolled into plates, structural sections, bars and strip. This takes about 40 minutes (compared with 10 hours in the open-hearth method) and may involve an initial charge of more than 350 tonnes [2].

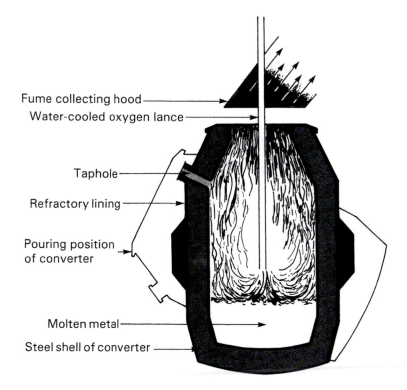

Figure 1.1   Basic oxygen steelmaking (BOS) process. (After reference 2.)

More recently the continuous casting process (CONCAST), in which the molten metal is poured directly into a casting machine to make the initial solid shape (known variously as slabs, blooms or billets depending on their size and shape), has been introduced for the production of structural sections. This eliminates many of the defects associated with production via the ingot route, leading to a better-quality final product. At a scale of production of a few millions of tonnes per mill per year this process is technically and economically sufficiently attractive for it to become the preferred process.

1.1.2   Rolling

At first sight it may appear strange that molten steel should first be cast into ingots which must then be reworked into usable shapes, rather than be cast immediately as plate, bar, etc. It is, of course, precisely this variety of products, as well as the practical difficulties associated with the casting of shapes such as wide thin sheets, that dictates the need for other processes. Moreover, the reheating, together with the actual mechanical working received during rolling, modifies the steel in such a way that its tensile strength is considerably enhanced. The most common treatment is hot rolling in which the steel is squeezed between a pair of rotating cylinders termed rolls. In this way the original ingots, weighing anything between 5 and 40 tonnes, are reduced in stages down to plate, strip (thin plate), sections, bars, wire, etc.

The sequence of operations involved in hot rolling [3] commences with the ingot being heated in a soaking pit for between two and eight hours. This is necessary in order to ensure that it attains an even temperature throughout (even when it 'solidifies' in the mould its size is such that the centre will still be molten). From here the ingots proceed to the primary rolling phase in which they are passed repeatedly through heavy rolls of the type shown in Figure 1.2. Each pass, of which there may be up to fifteen, reduces the thickness by as much as 50 mm. On emerging, the long slab or bloom has its ends cropped before passing through a second stage of rolling in the billet mill. The steel leaves here in the form of 10 m lengths of semi-finished material which are then inspected both visually and ultrasonically for surface and internal defects, such as cracks, blowholes and major slag inclusions. It is then reheated by passing through a series of furnaces until it reaches the final series of profiled rolls – so-called because, as shown in Figure 1.3, they operate on all four edges – which turn it into recognizable structural shapes. Final shaping of flat products (plate, sheet and strip) usually takes place in a four-high mill, in which the presence of the outer rolls reduces bending of the working rolls.

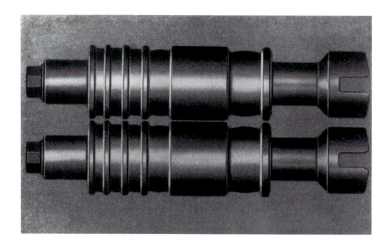

Figure 1.2   Rolls used for primary rolling of steel ingots. (*After reference 3.*)

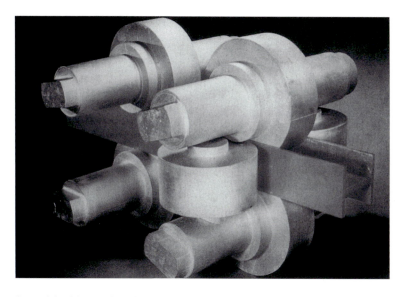

Figure 1.3   Model of profiled rolls used in final rolling of H-sections. (*After reference 3.*)

1.2   Properties of steel

Although the steelwork designer should be aware of all aspects [4] of the material he is using, his chief concern when making calculations which attempt to assess the load-carrying capacity of a particular member will normally be material strength. This property is usually measured in a tensile test in which a small coupon of material is pulled in a testing

machine until it fractures. Such tests also furnish useful information on
material stiffness and ductility. Guidance on the tensile testing of structural steel specimens is given in EN 10045, which covers items such as
specimen dimensions, testing speed and the proper interpretation of the
results. Figure 1.4 shows some typical tensile test specimens, while Figure
1.5 gives details of their recommended proportions.

The results of a tensile test are normally quoted in terms of a stress
strain curve for the material. A typical curve for structural mild steel is
shown in Figure 1.6 with an enlarged version of the most important, initial
portion being given in Figure 1.7.

When load is first applied the specimen responds initially in a linear
elastic fashion and obeys Hooke's law. Stress is directly proportional to
strain and removal of the load results in the strain falling to zero. The
slope of this straight-line portion is Young's modulus, E. As the strain is
increased a point is reached at which the curve tends to depart from lin-
earity. The stress at which this occurs is termed the 'limit of proportionality' σ_p and its presence is often difficult to detect. Further straining will
result in the steel yielding at a stress equal to the upper yield point σ_{yu}.
Once this stage has been reached the material no longer behaves elastically; even complete removal of the load will leave some permanent defor-

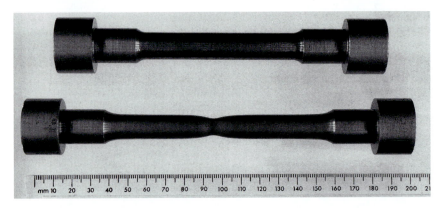

Figure 1.4   Typical tensile test specimens showing elongation immediately prior to
failure. (G.J. Davies)

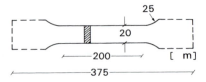

Figure 1.5   Typical dimensions of a rectangular cross-section tensile test specifically
according to EN 10045.

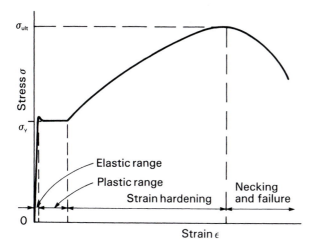

Figure 1.6 Typical stress-strain curve for structural mild steel obtained from a tensile test.

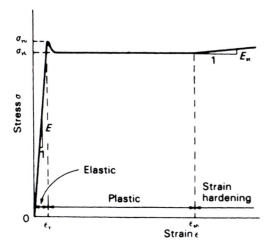

Figure 1.7 Enlarged, slightly idealized initial portion of the tensile stress-strain relationship for structural steel.

mation in the specimen. As the strain proceeds beyond ϵ_y so the stress drops slightly to the lower yield stress σ_{yL} (often called simply yield stress σ_y). The margin between σ_{yu} and σ_{yL} depends on the type of steel and also on the speed at which the test is conducted, with a typical value of the ratio σ_{yu}/σ_{yL} for normal structural steel being about 1.05–1.10, although higher values have been observed in tests involving particularly low rates of straining [5].

Tests at too high a rate may well result in a complete failure to observe an upper yield point [6]. Once the stress has dropped to σ_{yL} it remains sensibly constant for considerable increases of strain as shown in Figure 1.6. During this phase plastic flow of the material is taking place, the extent of which is a measure of the ductility of the material. Typical structural steels possess yield plateaus of at least ten times the strain at yield. Eventually yielding ceases and the stress starts to rise as the material strain hardens. The initial slope of this part of the curve is termed the strain hardening modulus E_{st}. It is much less steep than the elastic part, with E_{st}/E being typically between about 1/30 and 1/100 [5]. Eventually a maximum is reached on the stress axis; this corresponds to the ultimate tensile stress, σ_{ult}. Thereafter stress appears to decrease until fracture finally occurs. However, the real stress is actually still increasing; an apparent decrease is seen because the plotted quantity (often termed the 'engineering stress') is calculated using the original area whereas once σ_{ult} has been attained the actual area of the specimen decreases quite rapidly, a phenomenon known as 'necking'.

Although it is possible to conduct compressive tests on coupons this is complicated by the need to prevent the specimen buckling sideways. The results of such tests show the behaviour of most structural steels to be very similar in compression and tension, with the compressive yield stress being some 5% higher on average than the tensile value [7].

Ductility is measured by the percentage elongation, i.e. the increase in length divided by the original length measured over a standard gauge length (50 mm or 200 mm) obtained in the above test. Values as high as 20% of the original specimen length may be obtained. It is this property that enables small regions that are very highly stressed to yield, thereby relieving this concentration of stress without undue distress to the structure as a whole. Adequate ductility is also a prerequisite for the use of the plastic design methods described in Chapter 11.

1.2.1   Comments on yield stress

Reported values of the mechanical properties of the structural grades of steel used in the UK are listed in EN 10029 (plates), EN 10025 (sections) and EN 10210 and EN 10219 (hollow sections). Compliance with these is normally the responsibility of the steel producer, who will seek to ensure that this has been achieved through tests on samples taken from each batch of steel. The results of these tests are shown on the mill certificate. In cases where such tests reveal a shortcoming it is possible that the batch may be sold as a lower-quality grade, providing, of course, that it meets the minimum standards for that grade. Because of this it is sometimes possible for the user to find that in several respects his material possesses better properties than he expected. While this may be of significance to the researcher (high-strength material means that he will require higher loads for his laboratory tests) it should not worry the designer; indeed because designers normally use specified properties, only rarely calling for their own materials tests in the case of steel, it is something of which he will probably remain unaware.

Tensile tests performed by the manufacturer are frequently referred to as 'mill tests'. It is usual for them to be conducted at a fairly high rate of loading (a strain rate of 0.0025/s is mentioned in EN 10025). This is important because the yield stress of steel is strain-rate dependent [6], namely the results obtained from a tensile test will be influenced by the speed at which that test is conducted. Figure 1.8 illustrates this point.

By stopping the separation of the jaws of the testing machine so that the specimen is in effect being loaded at zero strain rate for a short period, it is possible to determine a minimum value of lower yield stress a few per cent below that which corresponds to straining on the yield plateau at a finite rate. This is termed the static yield stress and is illustrated in Figure 1.9. Because the majority of loads on civil engineering structures are usually considered to be of an essentially static nature, it is generally accepted that the static yield stress is the most appropriate basis for normal design calculations. Mill tests, however, tend to measure a higher, dynamic figure which, because of the procedure employed, will also contain some upper yield point effects [5]. It is therefore comforting to find that the average values of yield stress obtained from mill test results may be expected to lie significantly above the guaranteed minimum values of EN 10025. As an example, Figure 1.10 shows the results of tests on American ASTM A7 steel (broadly equivalent to S275) plotted as a frequency distribution. From the interpretation of these data given by McGuire [8] it would appear that only about 1% of mill test results fall below the specification value of 226 N/mm². However, if it is accepted that the static yield stress lies some 15% below the typical mill test value [9] then a mill test result of 260 N/mm² is necessary to ensure a static value of 226 N/mm². From Figure 1.10 it may be seen that some 40%

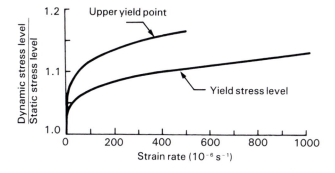

Figure 1.8 Effect of strain rate on upper yield point and yield stress of structural steel. (*After reference 5.*)

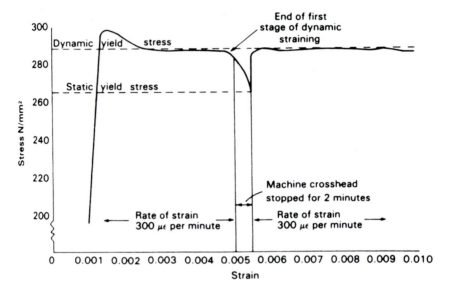

Figure 1.9   Definition of static yield stress from testing machine load–strain relationship.

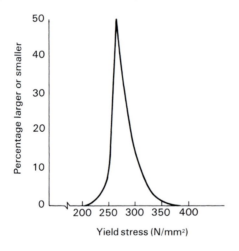

Figure 1.10   Variation of yield stress obtained from the results of 3974 mill tests on ASTM A7 steel. *(McGuire: Steel Structures, 1968. By permission of Prentice-Hall, Inc.)*

of samples fall below this figure. However, since the majority of these are not more than about 10% low, the net effect when averaged over a complete design is not likely to prove significant. Both the 15% difference between mill test results and the static yield stress and the shape of the distribution shown in Figure 1.10 have been confirmed by a Swedish study [5].

In the case of structural sections the difference between material strength as indicated by the mill test and the real, that is static, yield stress may also be influenced by the position from which the specimens are taken. For I-sections, EN 10025 normally requires these to be cut from the flange as shown in Figure 1.11. Since web material is thinner than that of the flanges it will tend to possess a slightly finer grain structure as a result of faster cooling after rolling. The importance of this to the structural designer lies in the fact that the yield stress will be higher [5, 9]. For a series of UB sections in S275 steel differences of up to 16% of flange yield strength have been observed [10]. In most situations it is the flanges of an I-section that contribute most to its load-carrying capacity. For instance, even when it is used as a tie, most of the load will be carried by the flanges simply because most of the area is concentrated in the flanges, while members in bending derive virtually all their strength from the contribution of the flanges.

It is clear from the above discussion that the structural designer must be careful in selecting the appropriate value of material strength for use in his calculations. For the usual structural steels according to EN 10025, a set of values of design strength p_y is given in *Table 9* of BS 5950: Part 1. These values differentiate between different grades, thicknesses and types of section. They are based upon the specified minimum yield stresses given in Table 4 of EN 10025, suitably adjusted by a partial safety factor on material strength (for explanation of partial safety factors see Chapter 2) so as to allow for the effects of all of the factors discussed above.

1.2.2   Residual stresses

Figure 1.7 shows that the mechanical strain at yield for structural steel is of the order of 1%; this is approximately the same as the expansion produced

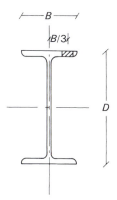

Figure 1.11   Location of tensile specimen in a steel I-section according to EN 10025.

by placing a piece of steel in boiling water, i.e. increasing its temperature to 100 °C, Much higher temperatures, typically 600–700 °C, are involved in the rolling of steel, while members fabricated by welding (possibly using material that has previously been flame cut) will be subject to a further application of heat. Moreover, this heating will be applied locally to selected parts of the cross-section. Cooling of the heated material will always take place unevenly, even for the hot-rolled member placed on the cooling bed after rolling, for which air will reach the extremities, such as the flange tips of an I-section, more readily.

The net result of these processes of uneven heating and cooling is that structural members will normally contain residual stresses. Although these may be removed by subsequent reheating and slow cooling the process is expensive and is limited to special components like pressure vessels, for which the presence of residual stresses is known to be particularly unwelcome. As a general rule those parts of the section which cool first will be left in residual compression, while those that cool more slowly will contain residual tensile stresses. The region adjacent to a weld will normally be stressed up to yield in tension with balancing residual compression elsewhere in the section. Figure 1.12 illustrates typical patterns of residual stress for a rolled I-section and a welded box.

Because residual stresses must themselves be in equilibrium, their effect on structural behaviour is limited; the most important consequence for statically loaded structures is to cause the member to behave as if it possesses

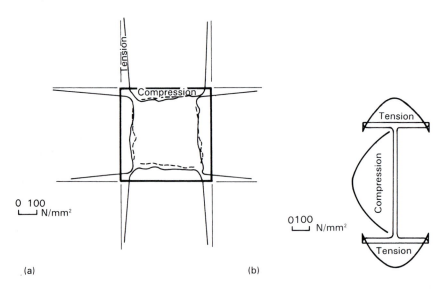

Figure 1.12   Typical measured patterns of residual stress in structural sections. (a) 400 × 400 × 5 mm corner welded box section [11]. (b) 250 × 146 UB rolled I-section [12].

a non-uniform distribution of yield stress over its cross-section. This is particularly important for compression members, for which those regions containing residual compression yield at loads producing an applied stress of less than σ_y. Members in bending also yield early and therefore tend to deflect more [13]. The presence of residual stresses also tends to lessen a member's resistance to the growth of cracks, whether this occurs in a stable manner due to the action of fluctuating loads (fatigue), or in an unstable fashion by the process known as brittle fracture [8].

1.3   Composition toughness and grades

Structural steels contain very small quantities of a number of elements, each of which has some influence on the physical properties of the steel. Most important of these is carbon; an increase in carbon content causes increases in both strength and hardness but at the expense of both ductility and toughness. Details of the required chemical compositions for all UK grades of structural steel are given in EN 10025.

Chemical composition also affects a steel's suitability for welding, a property known as weldability. Because welding is so extensively employed in the fabrication of structural steelwork it is important that the steel used be capable of being welded without the need for special, and therefore expensive, welding procedures. A measure of weldability is the so-called carbon equivalent, C.E. defined as

$$\text{C.E.} = \text{C} + \frac{\text{Mn}}{6} + \frac{\text{Cr} + \text{Mo} + \text{V}}{5} + \frac{\text{Ni} + \text{Cu}}{15}$$

in which each symbol refers to the proportion by weight of that particular element. Table 3 of EN 10025 gives values of C.E. of about 0.20%. Low values of C.E. imply good weldability. For details of the appropriate welding techniques, for example metal arc, submerged arc, etc., references should be made to BS EN 1011–1 and BS EN 1011–2. Readers wishing to acquaint themselves with the basic features of the welding of structural steel should consult reference [14].

Structural steel is available in the UK in three main grades: S275, S355 and S460, where the figures denote the design strength. Each grade is available in a number of different qualities – reflecting changes in chemical composition and the exact method of production – that vary in terms of cost and ease of availability. Generally speaking, better properties in terms of ease of welding (weldability) and greater resistance to impact (toughness) involve higher costs and lesser availability. The main structural grades are S275 (mild steel) and S355, with S355 being the principal grade for bridge work and being increasingly used in place of S275 for major structural members for buildings. The highest

The exposed structural framing of the Hong Kong and Shanghai Bank.

strength S460 for which σ_y is of the order of 440 N/mm², approaching twice that of S275, is rarely used. Present pricing policy is such that S355 costs about 10–15% more than S275 with the price of S460 being a further 25% higher.

Toughness is necessary in structural steel in order to avoid the phenomenon known as brittle fracture. This can cause complete failure by the very fast propagation of a small crack, often in regions of comparatively low stress. Much has been written about brittle fracture since the first failure was identified in 1886 [8]. An account of some of the subsequent failures is given by McGuire [8], who also describes the metallurgical process involved. From the designer's point of view the most satisfactory way of dealing with brittle fracture is to reduce the likelihood of its occurrence by a sensible choice of material. Providing the structure will not be subject to combinations of situations which are conducive to brittle fracture, such as low temperature, thick plates with mutually perpendicular welds stressed in the through-thickness direction and fast rates of loading, then this is not too difficult.

The approach taken by BS 5950: Part 1 is that brittle fracture is unlikely for routine applications of structural steelwork in the United Kingdom. Thus *Cl. 2.4.4* directs the reader to a table of 'safe' maximum material thicknesses. For potentially critical situations such as welding details which induce a high degree of restraint, the designer is advised to seek specialist guidance. Since the method by which this may be obtained is not specified, the onus rests with the designer to use his judgement and experience backed up by the advice of a materials specialist if the circumstances are thought to warrant it [15].

Specific guidance on the selection of suitable steel grades for each of the main types of structural section in terms of maximum plate thickness for different operating conditions is provided in *Cl 2.4.4*. This categorizes the choice according to the key influences:

Minimum service temperature
Thickness
Steel grade
Type of detail
Stress level
Strain level or strain rate

Normal operating temperatures are assumed not to fall below between −5 °C for internal steelwork and −15 °C for external steelwork, although data are provided down to −45 °C.

The basis of this information is that the steel exhibits sufficient energy absorption when subject to a Charpy vee-notch impact test [15]. This is a standard material test in which small bars containing a notch are fractured by a heavy pendulum, the energy required being determined from the swing of the pendulum. Results are normally quoted as Charpy values C_v. Since they are currently affected by temperature, C_v values must be related to the testing temperature and a figure of −5 °C is often taken as representing the minimum service temperature. More detailed information on the significance of Charpy test values and their relationship with true fracture toughness, as indicated by the application of the recently developed technique of fracture mechanics, is given by Burdekin [16] in a paper describing the basis for the toughness requirements for bridge steel in the UK.

1.4   Fatigue

In structures subject to a very large number of cycles of fluctuating loud, typically at least 100 000 load applications, failure may occur by the continued growth of cracks in the material at stresses well below those necessary to cause ordinary static yielding or collapse. Such behaviour is termed fatigue. Most civil engineering structures do not experience loads approaching their design load very frequently. Of course, there are certain exceptions, in crane girders, railway bridges, and offshore structures subject to wave loading. However, even ordinary wind loading does not normally provide sufficient repetitions unless the structure is susceptible to wind-induced oscillations. When it is realized that 100 000 cycles corresponds to ten applications a day for more than 25 years, it becomes clear that fatigue is unlikely to be a problem for ordinary building structures. It is more significant for bridges, although even here, since fatigue is largely dependent upon stress range, i.e. the difference between the maximum and minimum stresses experienced in service, many bridges will not receive sufficient applications of load heavy enough to produce the necessary large changes in stresses.

For the design of crane supporting structures BS 5950: Part 1 refers the

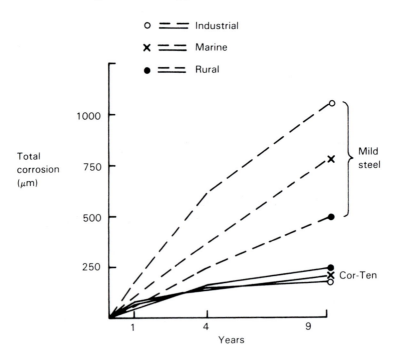

Figure 1.13   Comparative corrosion rates for mild steel and Cor-Ten steel in different environments. (After reference 23.)

engineer to BS 2573; for more general guidance the fatigue section of the bridge code [17] may be consulted. Readers wishing to learn something of the mechanics of fatigue are referred to the relevant section in McGuire [8].

1.5   Corrosion and corrosion protection

Steel readily corrodes (rusts) in moist air. Aggressive environments such as smoke, soot, sea water, acid or alkaline vapours will hasten the process. In a bad industrial area the rate at which the surface is 'lost' may reach 0.075 mm/year, more if particularly harmful agents such as sulphur dioxide are present. Structural steelwork therefore needs to be properly protected [18]; guidance on this subject is provided in BS 5493.

The most common forms of protective treatment involve covering the exposed steel, either with paint or with a metallic coating, or possibly in the case of sheeting with a plastic coat. Concrete is not generally regarded as being capable of affording sufficient protection (except in the case of reinforcement).

Paint systems are described in BS 2015. Generally a zinc- or aluminium-based priming coat is applied first so as to provide a good foundation for the later finishing coats. Care is necessary when using certain paints on account of their toxic nature; they should not be sprayed, applied in confined spaces or used on material that will subsequently be welded or flame cut.

Metallic coatings include galvanizing and sheradizing (both of which use zinc), electroplating, which is mainly confined to small items like fasteners, and metal spraying using either zinc or aluminium. Information on each of these techniques is provided in the relevant British Standard [19–22].

A common requirement for all schemes is cleanliness of the surface before treatment. For structural steelwork this is normally achieved by blast cleaning in which small abrasive particles such as iron are directed at the object using either compressed air or an impeller. Fabricating shops often arrange for incoming material to pass through the shotblasting plant on entry to the shop.

One alternative to the use of protective treatments consists of using a special corrosion-resistant steel which rapidly forms its own protective layer of oxide film. As shown in Figure 1.13 this has the effect of reducing the corrosion rate to a negligible level after a few years. In Britain such materials are called 'weathering steels' [23], of which the best known is Cor-Ten. Originally developed by the United States Steel Corporation, this is now produced under licence in Britain by Corus. Designs using weathering steel clearly ought to exploit its particular properties; information on these is available [23].

It is important to appreciate that no coating is completely impermeable. Moreover, not surprisingly, there is a fair degree of correlation between the cost of a particular treatment and the degree of protection afforded by it. Therefore in common with most aspects of design the question of protection against corrosion is largely a matter of economics. To assist the designer, BS 5493 lists eight classes of environment (five external and three internal). Good designers will also try to 'design for prevention' by avoiding traps for dirt and moisture. A particularly useful presentation of the main aspects of corrosion protection for structural steelwork, covering corrosion prevention aspects of detailed design, surface treatment and protective systems is provided in ref 18, whilst further details are available in the ECSC guide [24]. A further discussion on minimizing the effects of corrosion is provided in reference [25], Chapter 35.

1.6   Fire protection of structural steelwork

Although steel is an incombustible material, Figure 1.14 shows how its strength may be reduced substantially by the action of high temperatures of the sort experienced in a major building fire. Moreover, because of its good thermal conductivity a bare steel beam may well assist in spreading a fire by igniting combustible material located beyond fire-resistant bulkheads. Therefore for most types of building the steelwork must be provided with some form of fire protection. Exceptions occur for single-storey structures isolated from any neighbouring buildings, some multistorey carparks and certain other 'zero-related' buildings which can be shown not to be affected adversely by the heat generated by a fire.

High Temperature Steel Properties

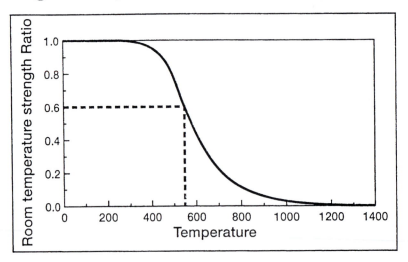

Figure 1.14   Effect of elevated temperature on the strength of structural steel.

In Britain the necessary requirements form part of the Building Regulations [26]. A useful explanation of how these relate to structural steelwork has been prepared by British Steel [27]. The essential point is that sufficient protection must be provided for the main skeleton of the building to stand up long enough for people inside to escape. Thus minimum periods ranging from 30 minutes for a small residential building to 4 hours for a large store, are specified.

Such protection is afforded normally by encasing the steelwork in a suitable fire-resistant medium. In the so-called 'traditional method', brickwork, blockwork or concrete encasement is used. Since a typical thickness might be 50 mm for two hours' protection such methods tend to be slow and labour-intensive. An alternative would be wrapping in metal mesh which could subsequently be covered with a suitable plaster. Lightweight methods involve spraying the steelwork with some form of proprietory product. These have the advantage of lightness (thereby contributing little to dead load) and are usually less bulky. They also permit easier modification to the structure, an important consideration when it is remembered that one of the main attractions of a steel structure is the relative ease with which it can be altered at a later date. Included in these modifications could, of course, be changes in required fire resistance; increasing the thickness of the sprayed protection is a relatively easy undertaking. A disadvantage, however, is that the actual spraying operation is messy and, because it involves an additional trade on site, complicates the construction programme. Thus the use of dry boards to enclose the steelwork is becoming increasingly popular. A brief comparison of the advantages and disadvantages of each of the main methods is provided in reference 27.

Whatever method is used, fire protection is a costly item and much attention is currently being given to ways of utilizing the inherent fire resistance of several forms of construction to reduce the need for added protection [27]. Some of the findings have been incorporated in the recently published Part 8 of BS 5950.

This provides the designer with guidance on the principles of designing fire resistance into his steel building – rather than simply accepting that fire protection will be required and then indicating suitable thicknesses of protection. As an example the thermal shielding effect of concrete slabs that form parts of particular forms of floor construction designed to exploit this property [27] is illustrated in Figure 1.15. This benefit is recognized by quoting much better resistance times than those given for either bare steel beams or beams with slabs located on their top flange.

Because of its increasing importance, the subject of design to provide adequate fire resistance is covered in some detail in Chapter 12.

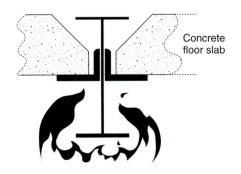

Slim Floor Construction

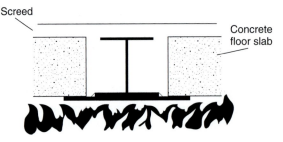

Figure 1.15   Shielding effect of concrete slab.

References

1 British Steel (1995) – *Construction versatility is our strength*, British Steel.
2 British Steel – *Making Steel*, British Steel Corporate Communications.
3 British Steel – *Shaping Steel*, British Steel Corporate Communications.
4 Owens, G.W. and Knowles, P.R., (1992) *Steel Designers' Manual*, 5th ed. Blackwell Scientific Press, pp. 199–225.
5 Alpsten, G.A. (1973) *Variation in Mechanical and Cross-sectional Properties of Steel*, Swedish Institute of Steel Construction, Publication No. 42.
6 Nagaraja Rao, N.R., Lohrmann, M. and Tall, L. (1966) 'Effect of strain rate on the yield stress of structural needs', *ASTM Journal of Materials*, 1(1), 241–64.
7 Transport and Road Research Laboratory (1977) 'Recommended Standard Practices for Structural Testing of Steel Models', TRRL Supplementary Report 254, Transport and Road Research Laboratory.
8 McGuire, W. (1968) *Steel Structures*, Prentice-Hall, Englewood Cliffs, New Jersey.
9 Beedle, L.S. and Tall, L. (1960) 'Basic column strength', *ASCE Journal of the Structural Division*, 86(ST7), 139–73.
10 Baker, M.J. (1972) 'Variability in the strength of structural steels – a study in structural safety. Part 1: Material variability', CIRIA, Technical Note 44, September.
11 Dwight, J.B., Chin, T.K. and Ratcliffe, A.T. (1968) 'Local buckling of thin-walled columns, Part 1: Effect of locked-in welding stress', CIRIA Research Report No 12, May.

12 Young B.W. (1972) 'Residual stresses in hot rolled members' Proceedings International Colloquium Column Strength, IABSE, Zurich, 25–38.

13 Tall, L. (1974) *Structural Steel Design*, Ronald Press, New York, NY.

14 Pratt, J.L. (1989), 'Introduction to the Welding of Structural Steelwork', SCI.

15 Rolfe, S.T. and Barsoum, J.M. (1977), *Fracture and Fatigue Control in Structures*, Prentice-Hall, Englewood Cliffs, new Jersey.

16 Burdekin, F.M. (1981) *Materials Aspect of BS 5400 Part 6, The Design of Steel Bridges*, edited by K.C. Rockey and H.R. Evans, Granada Publishing.

17 British Standards Institution (1980) 'BS 5400: Part 3, Steel Concrete and Composite Bridges; Part 10: Code of Practice for Fatigue', London.

18 British Steel (1996), *The Prevention of Corrosion on Structural Steelwork,* British Steel.

19 British Standards Institution (1971), 'BS 729, Hot Dip Galvanised Coatings on Iron and Steel Articles', London.

20 British Standards Institution (1988), 'BS 4921, Sheradised Coatings on Iron and Steel Articles', London.

21 British Standards Institution (1994), 'BS EN 22063, Metallic and Other Inorganic Coatings – Thermal Spraying, Aluminium and Their Alloys'.

22 British Standards Institution (1990), 'Method for specifying Electroplated Coatings of Cadmium and Zinc on Iron and Steel', London.

23 Chandler, K.C. and Kilcullen, M.B. (1973), 'Corrosion Characteristics of Weathering Steels', Technical Note 10/CAB/TN/73. Corrosion Advice Bureau, British Steel Corporation.

24 European Coal and Steel Community (1982) *Durability or Steel Structures.*

25 Owens, G.W. and Knowles, P.R. eds (1992) *Steel Designers' Manual*, 5th ed. Blackwell Scientific Press, London, pp. 998–1019.

26 Building Regulations 1991 (1992) Approved Document B, HMSO Building Standards (Scotland) Regulations 1990 (1990); Technical Standards, HMSO. Northern Ireland Building Regulations 1994 (1994), Technical Booklet E, HMSO.

27 British Steel (1997), *Fire Resistance of Steel Framed Buildings,* 1997 Edition, British Steel.

Chapter 2

The basis of structural design

Structural design is an all-embracing term, which is used to cover general aspects of the subject, for example the choice of a particular structural form and a particular material, through the series of increasingly narrower decisions that leads eventually to points of detail such as the size of bolt required in a particular connection. Progress through each of these stages usually involves treating the problem in an increasingly quantitative manner. Although this book is concerned largely with the more detailed end of the process as it applies to steel structures, the material of this chapter should provide the reader with a taste of the wider aspects of the subject. Since BS 5950 is written principally, but not exclusively, with steel building structures in mind, the text concentrates on examples drawn from that area. Readers wishing to gain a wider appreciation of steel structures should therefore consult some of the references given in the Bibliography at the end of this chapter.

2.1 Structural idealization

Once the decision has been taken to construct a particular building in steel a suitable structural system must be selected. Factors which might influence the choice include the following.

1 *The spans involved.* Special consideration is necessary if there is a requirement for long spans or large, clear floor areas.
2 *The vertical loading.* The presence of heavy point loads on floors or the need to accommodate cranes (*Cl. 2.2.3*).
3 *The horizontal loading.* Attention must be given to the way in which horizontal (wind) loading is to be resisted, for example by the framing itself (by providing rigid joints), by bracing acting with the framing or by means of an independent bracing system such as a set of shear walls. This aspect of design is of particular importance for very tall buildings (*Cl. 2.4.2.3*).
4 *The services required.* These include water, electricity and gas and

Trusses, columns and plate girders at Heysham power station.

often nowadays significant computing facilities and/or air conditioning systems, and are usually accommodated under the floors. In situations where large volumes of services are needed, as in hospitals, special forms of flooring permitting easy incorporation of the necessary pipework and ducting may be necessary. For some forms of commercial office accommodation provision of services may account for up to 30% of the overall cost.

5   *The ground conditions.* Clearly the type of ground on which the building is to be erected will dictate the form of foundations that must be used (pad, raft, piled, etc.) and this in turn must be taken into consideration when selecting the superstructure (*Cl. 2.4.2.9*).

Other items which might enter the discussion are the ways in which the building must be erected, accommodation of temperature effects (*Cl. 2.3*) and (if the steelwork is to be visible to the users such as the inside of the roof of an exhibition hall) the appearance. BS 5950: Part 1 also requires steel-frame buildings to be tied together adequately and, in the case of multistorey buildings, to be capable of withstanding a limited amount of local damage without collapse (*Cl. 2.4.5*).

The way in which the designer decides to satisfy these requirements, several of which may well tend to conflict with one another, constitutes a difficult and frequently relatively neglected aspect of structural design. Its solution, which must draw heavily on experience of past satisfactory schemes, structural judgement, discussions with those other professions concerned with the design of the building as well as the client or user, knowledge of fabricating shop capabilities and erection techniques, etc., lies beyond the scope of this text. Wide reading of descriptions of actual projects (case studies) [1, 2], discussions with practising engineers, visits to fabricating shops and construction sites as well as a clear appreciation of structural behaviour all form part of the necessary educational process. In comparison with this aspect of design the actual proportioning of the members, detailed design of the connections, etc. is normally much more straightforward. However, a proper understanding of the more limited task is necessary before an engineer is competent to tackle the problem in its wider sense. Even when this stage has been reached greater experience and career advancement will cause the engineer to reconsider his definition of structural design as the boundaries of his involvement become wider.

The majority of steel buildings fit within one of the categories listed in Table 2.1. Of these, bearing wall construction (Figure 2.1) in which the steel beams forming the roofs and floors bear directly on fairly substantial

Table 2.1   Broad categories of steel building construction

Type	Main use	Main considerations in design
Bearing wall	Low rise, lightly loaded	Structural design of steelwork is normally straightforward
Steel frame	Wide variety of types and size of building	'Simple construction' or 'continuous construction' depending on joint type used
Long span	Coverage of large column-free areas	Special forms of 'beam' may be required to span the required distances
High rise	Tall buildings, i.e. more than 20 storeys	Resistance to lateral forces due to wind load

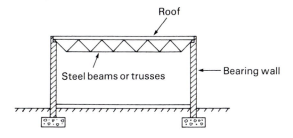

Figure 2.1   Bearing-wall construction.

walls (usually constructed of brick or concrete blocks but sometimes of plain or reinforced concrete), is usually limited to low-rise, lightly loaded buildings such as schools.

A steel framework of beams and columns such as that shown in Figure 2.2 is much more common nowadays. Great versatility is possible, permitting this form of construction to be used for small, simple low-rise buildings as well as for much more complicated multistorey buildings. Depending on the type of beam-to-column joints employed, such systems are considered either as 'simple construction' (*Cl. 2.1.2.2*) or as 'continuous construction' (*Cl. 2.1.2.3*). For the former, rotation of the beams relative to the columns is assumed to be possible so that beams may be designed as simply supported with columns required to carry only those moments produced by the eccentricity of the beam reactions (see Figure 2.3). Relatively simple connections may be used to transmit shear and these can usually be bolted up in the field without undue difficulty. Continuous construction (also called 'rigid frames') assumes sufficient rigidity in

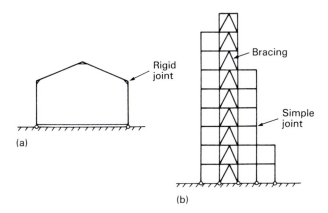

Figure 2.2   Beam and column construction: (a) portal frame – continuous construction; (b) multistorey frame – simple construction.

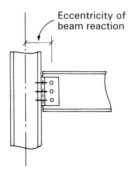

Figure 2.3   Column moment due to eccentricity of beam reaction.

the beam-column connections to maintain virtually unchanged the original angle between those two members when the structure is loaded. Such connections naturally involve additional fabrication and probably higher erection costs but the greater rigidity produced in the structure due to its ability to develop flexural action may well compensate in terms of reduced member sizes and the elimination of bracing. This form of construction is very popular for low-rise industrial buildings of the type shown in Figure 2.2(a).

One very significant difference in the approach to the design of these two types of framing is that because simple construction is effectively statically determinate all members can be designed more or less in isolation in a single pass through the structure, whereas the interactions between adjacent members present in continuous construction necessitates the consideration of at least a group of interconnected members. Since such subframes are statically indeterminate several cycles of design are often necessary.

For long-span construction, for example roofs, the floor directly over a hotel ballroom, etc., normal rolled sections may not have sufficient depth to act as beams. In such cases they may be replaced by plate girders or trusses. Coverage of very large areas may require the use of space frames, arches or even cable-suspended roofs. Detailed consideration of these more exotic forms of construction is beyond the scope of this text and the interested reader is referred to the Bibliography.

For tall structures such as buildings of more than about 20 storeys depending on circumstances, microwave towers, etc., considerations of resistance to lateral wind loading tend to dominate the design thinking. Figure 2.4 illustrates the two basic mechanisms for providing sway stiffness in a steel-frame structure; either it can be braced, possibly using the internal walls, lift shafts, etc., in which case adequate stiffness may be possible using main frames of 'simple construction', or sway may be resisted by the inherent bending stiffness of rigid frame action. Various special

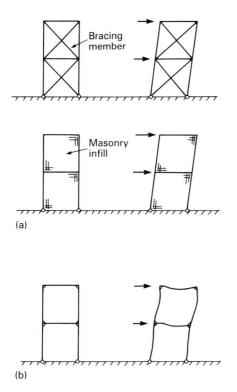

Figure 2.4   Basic methods of providing sway resistance in a steel frame: (a) bracing in simple construction; (b) rigid-frame action in continuous construction.

systems have evolved to permit the construction of the 70–110 storey buildings that currently represent the world's tallest.

2.2   Structural codes

Much of the detailed information necessary for the design of steel structures is provided in codes of practice. In the context of this book the most important of these is BS 5950: *The Structural Use of Steelwork in Building, Part 1: Code of Practice for Design in Simple and Continuous Construction*. Chapters 3–7, each of which deals with the design of structural elements, will make frequent reference to the design procedures contained in that document. These cover items such as the relationship between strength and slenderness for a steel column, recommendations for the adequate spacing of holes for bolted joints and guidance on deflection limits. Whilst it is clearly necessary for the steelwork designer to be familiar with the provisions of this code, it is equally important that he uses it in an intelligent fashion. The code does not cover every aspect of steelwork design; many facets of the subject simply cannot be quantified in the manner necessary for codification, others are encountered so rarely that it is not considered necessary to lengthen the document by their inclusion, and some are properly left to textbooks on theory of structures.

A code of practice may therefore be regarded as a consensus of what is considered acceptable at the time it was written. Thus it contains a balance between accepted practice and recent research presented in such a way that the information should be of immediate use to the engineer in conducting his design. As such it is regarded more appropriately as an aid to design containing stress levels, design formulae and recommendations for good practice, rather than as a manual or textbook on design.

The full list of Parts of BS 5950 is given in Table 2.2. In addition to the Part 1, this book makes direct reference to Parts 3.1 and 5 (Chapter 9) and Part 8 (Chapter 12).

The steelwork designer will often need to refer to a number of other codes covering topics such as steel properties, welding of structural steelwork, properties of steel fasteners (bolts) and loads on structures as well as the other steelwork codes aimed specifically at bridges, masts and towers, offshore structures and steel silos. In certain cases he may find it useful to consult the codes of other countries [3].

2.3   Limit states and partial safety factors

Limit-states design simply provides the basic framework within which the performance of the structure can be assessed against various limiting conditions. When formulating procedures nowadays it is customary to do so in a way which recognizes the inherent variability of loads, materials, construction practices and approximations made in design; this usually involves the use of some concept of probability. The limiting conditions are normally grouped under two headings: ultimate or safety limit states

Table 2.2   BS5950: *The Structural Use of Steelwork in Building*

Part 1	Code of practice for design in simple and continuous construction: hot rolled sections (1990).
Part 2	Specification for materials, fabrication and erection: hot rolled sections (1992)
Part 3	Code of practice for design in composite construction.
	Part 3.1 Design of simple and continuous composite beams (1990).
Part 4	Code of practice for design of floors with profiled steel sheeting (1994).
Part 5	Code of practice for design of cold formed sections (1998).
Part 6	Code of practice for design of light gauge sheeting, decking and cladding (1995).
Part 7	Specification for materials and workmanship: cold-formed sections and sheeting (1991).
Part 8	Code of practice for fire protection of structural steelwork (1990).
Part 9	Code of practice for stressed skin design (1994).

and serviceability limit states. Table 2.3 lists those limit states which are usually considered relevant for structural steelwork. The attainment of one or more ultimate limit states (ULS) may be regarded as an inability to sustain any increase in load. Serviceability (SLS) checks against the need for remedial action or some other loss of utility. Thus ULS are conditions to be avoided whilst SLS could be considered as merely undesirable. Since a limit-states approach to design involves the use of a number of specialist terms, simple definitions of the more important of these are provided in Table 2.4. A more detailed discussion of these and other matters relating to the general limit-states philosophy is provided in reference [4].

Table 2.3   Limit states for structural steelwork

Ultimate (safety) limits – ULS	Serviceability limits – SLS
Overall loss of equilibrium (overturning)	Excessive deformation
Strength limits (general yielding, rupture, transformation into a mechanism, etc.)	Excessive vibration
	Corrosion
Elastic or plastic instability	
Fatigue (leading to fracture)	
Brittle fracture	

Table 2.4   Definition of basic limit-states terminology

Term	Definition
A limit state	A condition beyond which the structure would become less than completely fit for its intended use. If this happens, the structure is said to have entered a limit state.
The ultimate or safety limit state	Inability to sustain any increase in load.
The serviceability limit state	Loss of utility and/or requirement for remedial action
Characteristic loads	Those loads which have an acceptably small probability of not being exceeded during the lifetime of the structure.
The characteristic strength of a material	The specific strength below which not more than a small percentage (typically 5%) of the results of tests may be expected to fall.
Partial safety factors	The factors applied to characteristic loads, and properties of materials to take account of the probability of the loads being exceeded and the assessed design strength not being reached.
The design load or factored load	The characteristic load multiplied by the relevant partial factor.
The design strength	The characteristic strength divided by the appropriate partial safety factor for the material.

BS 5950 is not the first UK code to be based on this approach; it was preceded in 1972 by CP 110 (now revised as BS 8110), the concrete code. Moreover, BS 5400, the bridge code, including Part 3 relating to the design of steel bridges, was prepared at much the same time as BS 5950, although it was actually published a few years previously. In other parts of the world limit-states steelwork codes are gradually appearing, with the first of these having been published in Canada as long ago as 1974. In the UK limit-states versions of the code for construction in aluminium and masonry are available; that for timber uses an alternative format. Thus BS 5950 simply reflects the trend towards the general introduction of this more rational approach to structural design that is taking place for all the major construction materials on a worldwide basis. A particularly important example of this is the production within the European Economic Community of Eurocodes. These are intended to fulfil a similar role within the EU as is done at present by national codes within individual member countries. At the time of writing draft versions of EC3 for steel structures and EC4 dealing with composite construction are available for trial use and are in the process of being converted into their final forms. Both documents reflect up-to-date technical thinking, harmonized so as to be acceptable to all the potential users. In time it is confidently expected that Eurocodes will replace national codes as the everyday working documents of designers.

Design for the ULS may conveniently be explained with reference to the type of diagram shown as Figure 2.5. This compares the strengths R of a number of nominally identical structures with the load spectrum Q that might be expected to occur during the lifetime of those structures. The fact that both quantities appear not as single vertical lines but as curves, termed frequency distributions, is in recognition of the variability not only

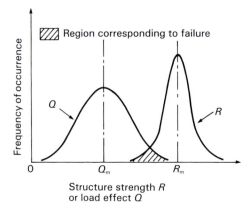

Figure 2.5   Pictorial representation of the variability of loads and strengths.

of the loads experienced by a structure but also of the factors which influence the strength of the structure. Thus the load curve is broad, reflecting
the variability of loading on a building structure, while the greater degree
of control over its strength leads to a narrower strength curve. A simple
illustration of the variability of structure strength R is provided by the data
given in Figures 2.6–2.8. These show how the naturally occurring variations in cross-sectional area and material strength of Figures 2.6 and 2.7
(together with various other properties not illustrated) contributed to the
spread of strengths shown in Figure 2.8 when the beams were tested.

The shape of both the load and the strength curves of Figure 2.5 will
always be such that some overlap will be present; this corresponds to a
failure. Good design consists of so proportioning the structure that this
area corresponds to an acceptably small probability (say 1 in 100 000). In

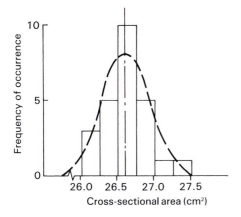

Figure 2.6   Variation in cross-sectional area of steel beam sections.

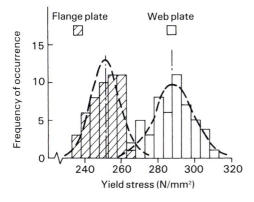

Figure 2.7   Variation in strengths of the material of steel I-sections.

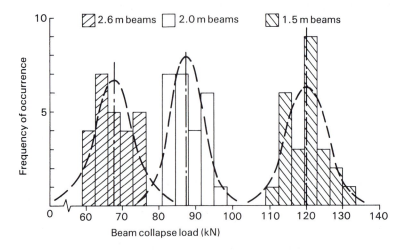

Figure 2.8   Strengths of three sets of 25 nominally identical steel beams.

traditional allowable stress design this is achieved by scaling down the strength side of the design equation using a factor of safety γ_e as indicated in Figure 2.9, while ultimate strength design compares actual structural strengths with the effects of factored-up loading by using a load factor γ_p. Figure 2.9 shows how limit-states design at the ULS employs separate factors on loading (γ_p) and strength (γ_m) in an attempt to cater for the different amounts of variability associated with these. Moreover, it is customary to break down the factors on each side into a number of partial safety factors, each of which reflects the degree of confidence in the particular contributing effect. Thus for a steel bridge for which the dead weight of the steelwork might be expected to be capable of more accurate assessment than the live loading due to traffic, the former will have a smaller

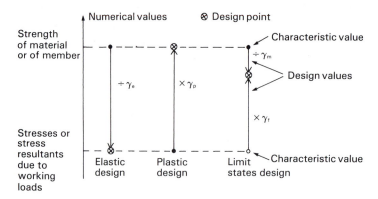

Figure 2.9   Level at which design calculations are conducted for different approaches.

partial factor associated with it than the latter. Typical figures might be 1.05 and 1.50 respectively. An internationally agreed list of γ-factors, as they are called, is available [4]. Actual numerical values are, however, usually decided upon in the code-drafting committees of an individual country.

BS 5950: Part 1 deliberately adopts a very simple interpretation of the partial safety factor concept in using only three separate γ-factors.

1 *Variability of loading* γ_l. Loads may be greater than expected; also loads used to counteract overturning may be less than intended.
2 *Variability of material strength* γ_m. The strength of the material in the actual structure may vary from the strength used in calculations.
3 *Variability of structural performance* γ_p. The structure may not be as strong as assumed in the design because of variations in the dimensions of members, variability of workmanship and differences between the simplified idealizations necessary for analysis and the actual behaviour of the real structure.

A value of 1.2 has been adopted for γ_p which, when multiplied by the values selected for γ_l, leads to the values of γ_f to be applied to the loading given in Table 2.5. Values of γ_m have been incorporated directly into the design strengths given. Thus the designer needs to use only the γ_f-values of Table 2.5 in his calculations. The different numerical values shown here are intended to provide approximatley equal margins of safety under each form of loading.

2.4 Loading

Assessment of the design loads for a structure consists of identifying the forces due to both natural and man-made effects which that structure must withstand and then assigning suitable values to them. Frequently several different forms of loading must be considered, acting either singly or in combination, although in some cases the most unfavourable situation might be easily identifiable. For buildings the usual forms of loading

Table 2.5   Values of γ_f to be applied to the loading – see *Table 2* of BS 5950: Part 1 for full list.

Load type	Value of γ_f
Dead (maximum)	1.4
Dead (minimum)	1.0
Imposed (in the absence of wind)	1.6
Wind (acting with dead load only)	1.4
Wind and imposed (acting in combination)	1.2

include dead load, live load, wind load, loads due to temperature effects and, in certain parts of the world, earthquake load. Other types of structure will each have their own special forms of loading, for example vehicle loading on highway bridges, fluid pressure inside storage tanks, and wave loading on marine structures.

When assessing the loads acting on a structure it is usually necessary to make reference to the appropriate codes of practice. Basic data on dead, live and wind loads for buildings in the UK are given in BS 6399 [5] with more specialised information on matters such as the loads produced by cranes in industrial buildings (workshops, steel plants, etc.) being provided elsewhere [6]. For bridges and other special forms of structure the necessary loading data are normally provided in the code of practice appropriate to that type of structure [7, 8].

Determination of the dead load of a structure requires the estimation of the weight of the structure together with its associated 'non-structural' components. Thus, in addition to the bare steelwork (which strictly speaking should include items such as bolts and weld metal), the weights of floor slabs, partition walls, ceilings, plaster finishes and services (cable ducts, water pipes, etc.) must all be calculated. Since certain of these will not be known until after at least a tentative design is available, designers normally use approximations based on experience for their initial calculations. As an example, the weight of the steelwork in a light roof truss may be assumed as $50 \, \text{kg/m}^2$. When the design is complete the actual dead load should be calculated; if it is significantly different from the assumed value then some modification of the design may be necessary. For the majority of steel buildings the weight of the actual steelwork will be less than 30% of the total dead load, so that quite large inaccuracies in its original assessment are unlikely to result in significant redesign.

The basis for estimation of live load is observation and measurement [9]. Live load in buildings covers items such as occupancy by people, office floor loadings, movable equipment within the building, and machinery. Clearly different values will be appropriate for different forms of building – domestic, offices, warehouses, etc. The effects of snow, ice and hydrostatic pressure are normally included in this category.

Although the load produced on a structure by the action of the wind is really a dynamic effect, it is normal practice for most types of structure to treat this as an equivalent static load. Therefore, starting from the basic wind speed for the geographical location under consideration, suitably corrected to allow for the effects of factors such as topography, ground roughness and length of exposure to the wind, a dynamic pressure is determined. This is then converted into a force on the surface of the structure using pressure or force coefficients which depend on the building's shape. For some surfaces the final effect may well be to produce a negative suction force. The information contained in reference [5] is limited to the more usual building shapes; for larger and more complex arrangements the designer may require a model of his structure to be tested in a wind tunnel. Very tall buildings, high masts and suspension bridges often fall into this category.

The designer must also decide whether allowance is necessary for any temperature effects. These include expansion or contraction due to temperature difference, such as between the sunny and shaded parts of a bridge, as well as shrinkage and creep, as with concrete slabs.

References

1   Institution of Structural Engineers, *Case Studies Nos. 1–7.*
2   British Steel (1985) *Structural Steel Design Teaching Project.*
3   British Constructional Steelwork Association (1983) *International Steel Handbook*, London.
4   Construction Industry Research and Information Association (1977) *Rationalisation of Safety and Serviceability Factors in Structural Codes*, CIRIA Report 63, July.
5   British Standards Institution (1996, 1997, 1998) BS 6399: Parts 1, 2 and 3, *Design Loading for Buildings*, London.
6   British Standards Institution (1983) BS 2573: Part 1, *Rules for Design of Cranes Part 1. Specification for Classification, Stress Calculations and Design Criteria for Structures*, London.
7   British Standards Institution (1978) BS 5400: Part 2, *Steel, Concrete and Composite Bridges*, London.
8   British Standards Institution (1986) BS 8100 *Lattice Towers and Masts: Part 1 Code of Practice for Loading*, London.
9   Mitchell, G.R. and Woodgate, R.W. (1971) *Floor Loadings in Office Buildings – the Results of a Survey*, BRS Current Paper 3/71, January.

Bibliography

Adams, P.F., Kulak, G.L. and Gilmor, M. (1990) *Limit States Design in Structural Steel*, 4th edn, Canadian Institute of Steel Construction.
Ballio, G. and Mazzolani, F.M. (1983) *Theory and Design of Steel Structures*, Chapman and Hall, London.
Bancroft, J. and Rogers, P. (1987) *Structural Steel Classics 1906–1986*, British Steel.
Bresler, B. and Lin, T.Y. (1964) *Design of Steel Structures*, Wiley, New York.
British Steel Corporation (1980) *Construction Guide*, BSC Sections, Redcar.
Clarke, A.B. and Coverman, S.M. (1987) *Structural Steelwork: Limit State Design*, Chapman and Hall, London.
Dowling, P.J., Knowles, P. and Owens, G.W. (1988) *Structural Steel Design*, Butterworths, London.
Gaylord, E.H. and Gaylord, C.N. (1972) *Design of Steel Structures*, 2nd edn, McGraw-Hill-Kogakusha, Tokyo.
Gimsing, N.J. (1983) *Cable Supported Bridges, Concept and Design,* Wiley-Interscience, Chichester.
Hart, F., Henn, W. and Sontag, H. (1985) *Multi-Storey Buildings in Steel*, 2nd edn, BSP Professional Books, Oxford.

Hayward, A.C.G. and Weare, F.E. (1988) *Steel Detailer's Manual*, BSP Professional Books. Oxford.

Heins, C.P. and Firmage, D.A. (1979) *Design of Modern Steel Highway Bridges*, Wiley-Interscience, New York.

Makowski, Z.S. (1965) *Steel Space Structures*, Michael Joseph, London.

McGuire, W. (1968) *Steel Structures*, Prentice-Hall, Englewood Cliffs, New Jersey.

Modern Steel Construction in Europe (1963) Elsevier, Amsterdam.

Morris, L.J. and Plum, D.R. (1988) *Structural Steelwork Design to BS 5950*, Longman, Harlow.

O'Conner, C. (1971) *Design of Bridge Superstructures*, Wiley-Interscience, New York.

Podolny, W. and Scalzi, J.B. (1976) *Construction and Design of Cable-Stayed Bridges*, Wiley, New York.

Salmon, C.E. and Johnson, J.E. (1980) *Steel Structures*, 2nd edn, Harper and Row, New York.

Trahair, N.S. and Bradford, M.A. (1998) *The Behaviour and Design of Steel Structures, to AS 4100*, 3rd edn, E & FN Spon, London.

Troitsky, M.S. (1988) *Cable Stayed Bridges, Theory and Design*, 2nd edn, Blackwell Scientific Publications, Oxford.

Woolcock, S.T., Kitipornchai, S. and Bradford, M.A. (1999) *Limit State Design of Portal Frame Buildings*, 3rd edn, AISC, Australia.

Papers on steel construction appear in numerous professional journals from time to time; the following journals specialize in the subject:

Modern Steel Construction. Published six times a year by the American Institute of Steel Construction, contains papers describing US projects.

New Steel Construction. Published six times a year, the *Journal of the Steel Construction Institute and the British Constructional Steelwork Association* contains papers describing both technical advances and new projects.

Chapter 3

Tension members

Tension members are used quite frequently in a variety of steel structures; some of these uses are illustrated in Figure 3.1. Depending principally upon the magnitude of the load to be carried and the type of interconnection to be used between members, any of the structural sections shown in Figure 3.2 may be suitable. Although the major design consideration will be the provision of adequate tensile strength, some limitation on slender-

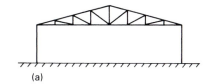

(a)

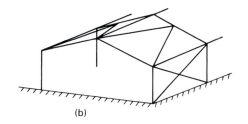

(b)

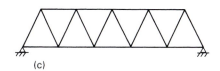

(c)

Figure 3.1   Structures containing tension members: (a) roof truss; (b) bracing for a portal frame building; (c) bridge truss.

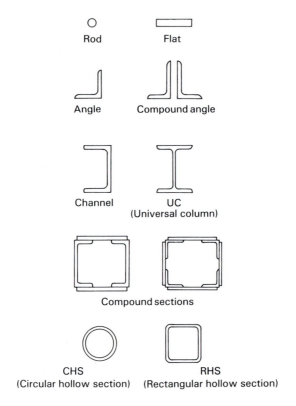

Figure 3.2   Examples of tension members.

ness is usually also necessary in order to eliminate possible problems due to excessive sag under self-weight [1], flutter due to wind loads or vibration caused by moving loads. For this reason rods or flats are of limited use, especially if required to act in compression due to reversal of load. When used as diagonal bracing, rods may be pretensioned so as to reduce their self-weight deflections [1].

Angles, used either singly or in pairs placed back to back, are suitable for many applications; two of the more visible examples are in small to medium roof trusses or in transmission towers. When heavy loads have to be carried over long spans, as in a truss bridge, then large rolled sections, possibly acting in combination, may be necessary. Tubes, either circular or rectangular, may be used as bracing or as the main members in trusses or space frames; care is necessary in deciding upon the jointing arrangements [2].

3.1   Behaviour of members in tension

The design of a member subjected to a tensile force is probably the most straightforward of all structural design problems. Essentially it consists of ensuring that the cross-sectional area of material provided is at least adequate to resist the applied load. Most students will have witnessed the standard laboratory test described in Section 1.3. The behaviour of a tension member is in many respects very similar, the most important difference being that the member will be attached to other parts of the structure. Whatever method of jointing is employed, whether bolting or welding, it will influence the manner in which load is transferred into the member. In the case of fastening by mechanical means the presence of the holes will also have a direct effect on the member's strength.

This problem is usually discussed in terms of gross and net sections. The former is simply the original cross-section while the net section is usually defined as the reduced section at a line of holes, i.e. gross section minus allowance for holes. The effect of a hole in a tension member amounts to more than simply the absence of some material. In the immediate vicinity of the hole a stress concentration will be present and this will itself be affected by the localized force applied by the fastener. However, because of the ductility possessed by structural steels, it is normal in design to neglect these other effects and to calculate the net section simply by subtracting the area of the hole(s). In doing this it must be remembered that most types of bolt (see Section 7.1.1) are used in clearance holes, where the hole is made slightly larger than the bolt diameter, usually 2 mm larger for bolt diameters up to 24 mm.

Since removal of material may be expected to have a weakening effect one might expect that failure would normally occur at the smallest net section, i.e. across the line of holes AA in Figure 3.3. However, because it is desirable that failure occur in a ductile rather than a brittle manner, it is usual to try to ensure that the gross section yields before the ultimate tensile strength of the net section is reached [3]. This greatly increases the amount of deformation that the member can sustain and consequently gives a better indication of impending failure.

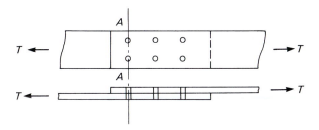

Figure 3.3   Tearing at a line of holes.

Even for a connection between two flat bars of the type shown as Figure 3.3 some eccentricity of the line of action of the tension T will be present. However, providing this is small, the resulting bending effects will be such that their influence on the member's ultimate strength may be neglected. Thus BS 5950: Part 1 allows certain types of tension member to be designed for tension only, makes safe approximate allowances in other cases, and only occasionally requires the explicit consideration of bending effects.

3.2   Basic design approach

3.2.1   Effective section

By ensuring that the ratio of net area A_n to gross section A exceeds the ratio of yield strength Y_s to ultimate tensile strength U_s, BS 5950: Part 1

Europe's highest building: the Jungfrau hotel.

effectively allows design to be based on the condition of yield of the gross section. Thus the effective area at a connection A_e is defined in *Cl. 3.4.3* as K_e times the net section, where K_e adopts the values 1.2, 1.1 and 1.0 for steel grades S275, S355 and S460 respectively, with the limitation that the effective area cannot exceed the gross section. The tension capacity P_t of the member is therefore given by

$$P_t = A_e p_y \tag{3.1}$$

in which p_y is the design strength of the steel obtained from *Table 9*. Reference to this table shows that it differentiates between the three basic grades of structural steel. It also allows for the gradual reduction in material strength that results from the use of thicker material.

3.2.2   Net section

Where holes are arranged in parallel rows at right angles to the member axis as shown in Figure 3.3 the net section is obtained by subtracting the maximum sum of the hole areas across any cross-section from the gross area, i.e.

$$A_n = A_g - \Sigma td \tag{3.2}$$

in which t = plate thickness and d = hole diameter.

Example 3.1

A flat bar 200 mm wide $\times$ 25 mm thick is to be used as a tie. Erection considerations require that the bar be constructed from two lengths connected together with a lap splice using six M20 bolts as shown in Figure 3.3. Calculate the tensile strength of the bar assuming steel of design strength 265 N/mm².

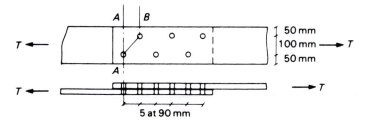

Figure 3.4   Zig-zag failure mode for staggered holes.

Solution

Hole clearance = 2 mm

Gross section = 200 × 25 = 5000 mm²

From equation (3.2), Net area AA = 200 × 25 − 2 × 22 × 25 = 3900 mm²

From *Cl. 3.4.3* for S275 steel effective area = 1.2 × 3900 = 4680 mm², which is less than the gross area.

From equation (3.1), P_t = 265 × 4680 N = <u>1240 kN</u>

Inspection of Example 3.1 reveals that the effective section is some 94% of the gross section. Over most of the member's length the section is therefore overdesigned, i.e. its capacity exceeds the required strength. This will usually be the case when parallel rows of holes are present. However, it is possible to reduce or even to eliminate this overdesign by staggering the holes as shown in Figure 3.4. This introduces the possibility of failure occurring at either of the two net sections AA across the plate or AB in a zig-zag. Both sections should normally be checked.

Calculations of the net section at a line of staggered holes is covered in *Cl. 3.4.4.3*. This makes some allowance for the slightly increased strength corresponding to the zig-zag mode, by reducing the amount of materials considered as ineffective to the total hole area along the section less a factor, to give

$$A_e = A_n + \frac{S_p^2 t}{4g}$$
(3.3)

in which t = thickness of plate and S_p and g are the staggered pitch and gauge as shown in Figure 3.5.

Example 3.2

Repeat Example 3.1 for the new arrangement of holes shown in Figure 3.4.

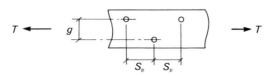

Figure 3.5   Definition of gauge *g* and staggered pitch *S*ₚ.

Solution
From equation (3.2), net section AA = 200 × 25 − 22 × 25 = 4450 mm²
From equation (3.3), net section AB

$$= 200 \times 25 - 2 \times 22 \times 25 + \frac{90^2 \times 25}{4 \times 100} = 4406 \text{ mm}^2$$

Minimum net section is AB and the effective area is therefore
1.2 × 4406 = 5287 mm² which exceeds the gross section.
  Take effective area as 5000 mm² and from *Cl. 4.6.1*
  Tensile strength = 265 × 5000 N = <u>1325 kN</u>

Thus staggering the holes results in a situation where design is governed
by the condition of yield of the gross section with no loss of efficiency.

3.3   Eccentric connection

Although it is usually regarded as 'good practice' to try to ensure that load
is transmitted into a tension member so that it acts along the member's
centroidal axis, this will not always be possible. One obvious example
would be a single angle for which the centroidal axis lies outside the cross-
section and connection to one or other leg would clearly introduce an
eccentricity. In other cases practical considerations of the geometrical
setting out of the joints in a truss will dictate that some eccentricity in the
line of action of the forces be introduced. For certain types of member,
however, the moments produced by these eccentricities are relatively
small and it is not actually necessary either to calculate them or to make
explicit allowance for them in design. Rather, the effective area may
be reduced slightly as shown in Figure 3.6 so that part of the member's

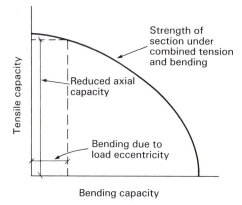

Figure 3.6   Use of reduced axial capacity to allow for interaction of tension and bending.

capacity which is not now being used to carry axial load is available to withstand the bending [4]. The justification for this approach is quite simply that, providing the correct sort of reduction in effective area is made, then it can be shown to provide good estimates of the strengths of single angles with either welded [4] or bolted [5] gusset plate connections on one leg.

Due to the eccentricity of load application the initial tendency as such a member is tested is for the gussets to deform so as to enable the line of action of the applied load to approach the centroidal axis of the angle, as illustrated in Figure 3.7. Thus at a load of about 50% of ultimate [4], while strains near the centre will be approximately uniform over the cross-section, sufficient bending will have occurred near the ends for yield of the attached leg to have started. This load may be determined approximately as that which just causes yield assuming the tension to act at the midplane of the gusset. Taking c as the distance between the centroid of the angle and the extreme fibre in contact with a gusset of thickness t, this gives a moment of $P(c + t/2)$. Rearranging for the value of P at which first yield occurs gives

$$P_y = A\sigma_y\left(\frac{1}{1 + Ac(c + t/2)/I}\right) \tag{3.4}$$

in which I = second moment of area of the angle about an axis parallel to the gusset.

Further loading will eventually produce full section yield in the central region, leading to large deformations until eventual failure by fracture of the angle in the connected region. The authors of both references [4] and [5] recommended that design be based on the load corresponding to the commencement of large deformations and suggested an expression of the form

$$P = \sigma_y(a_1 + fa_2) \tag{3.5}$$

in which a_1 = net area of connected leg
a_2 = gross area of outstanding leg
f   = factor to allow for bending effects
Thus the effective area A_e of the section is

$$A_e = a_1 + fa_2 \tag{3.6}$$

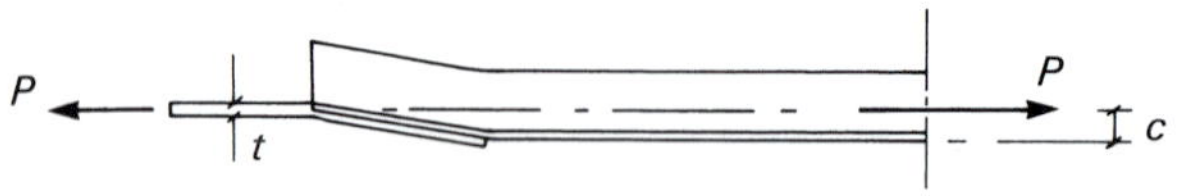

Figure 3.7   Bending of gusset to permit reorientation of load in angle connected to one leg.

Use of equation (3.5) was shown to provide quite consistent predictions of capacity.

3.3.1   Design approach

The method of allowing for eccentric connection when determining tension capacity P_T is explained in *Cl. 4.6.3*. For single angles connected through one leg only, single channels connected through the web or single tees connected through the flange P_T is obtained as:

$$P_T = p_y(A_e - 0.5a_2) \qquad \text{for bolted connections}$$

$$P_T = p_y(A_e - 0.3a_2) \qquad \text{for welded connections} \qquad (3.7)$$

in which $a_2 = A_g - a_1$
and a_1 = gross area of the connected element

When two identical parallel components are in contact back-to-back or are separated by a small gap with regular and frequent interconnection, equation (3.7) may still be used providing the factors 0.5 and 0.3 are replaced by 0.25 and 0.15 respectively when using bolts or welds.

Example 3.3

Determine the axial resistance of an $80 \times 60 \times 6$ mm, angle section when it is used as a tie, the end connection being a single row of M20 bolts through the longer leg. Assume steel of design strength $p_y = 275$ N/mm².

Solution
From section tables [6], $A = 80$ mm, $B = 60$ mm, $t = 6$ mm, Gross area of whole section $A_g = 808$ mm²

Gross area of connected leg $a_1 = 6 \times 80$
$$= 480 \text{ mm}^2$$

From equation (3.7) $a_2 = 808 - 480$
$$= 328 \text{ mm}^2$$

Noting from *Cl. 3.4.3* that $K_e = 1.2$,
$$A_e = 1.2\,(808 - 22 \times 6)$$
$$= 811 \text{ mm}^2 \text{ but} \leq 808 \text{ mm}^2$$

$\therefore$ use A_e $= 808$ mm²

From equation (3.7) $P_T = 275\,(808 - 0.5 \times 328)$
$$= \underline{177\text{kN}}$$

This compares with a member strength (no allowance for holes or eccentricity of 222 kN, i.e. a loss of 26%. This suggests that when selecting an initial trial section to check whether it will be adequate to resist the design load, a member whose area exceeds that given by (design load)/(design strength) by about 30% should be chosen. The 'extra capacity' should then be enough to balance the necessary allowances for holes and eccentricity. If welded end connections are to be used the margin should be reduced to about 15% since only eccentricity is involved. Clearly the exact amount that will be 'lost' depends on the relative areas of the connected and unconnected parts of the section.

Example 3.4

Select a suitable equal angle section to carry a tensile force of 240 kN assuming (a) a single row of M20 bolts, (b) welded end connections. Assume steel of design strength 355 N/mm².

Solution (a)
Approximate required area $= 1.3 \times 240 \times 10^3/355$
$$= 879 \text{ mm}^2$$

From section tables [6], nearest is $70 \times 70 \times 8$ mm which has an area of 1070 mm²

Gross area of connected leg $a_1 = 8 \times 70$
$$= 560 \text{ mm}^2$$

From equation (3.7) $a_2 = 1070 - 560$
$$= 510 \text{ mm}^2$$

Noting from *Cl. 3.4.3* that $K_e = 1.1$,
$$A_e = 1.1 \times (1070 - 22 \times 8)$$
$$= 983 \text{ mm}^2 \text{ but } \leq 1070 \text{ mm}^2$$

$$\therefore \quad \text{use } A_e = 983 \text{ mm}^2$$

From equation (3.7) $P_T = 355 (983 - 0.5 \times 510)$
$$= \underline{258 \text{ kN}} \text{ which is satisfactory}$$

(b) Approximate required area $= 1.15 \times 240 \times 10^3/355$
$$= 777 \text{ mm}^2$$

From section tables [6], nearest section is $70 \times 70 \times 6$ mm which has an area of 819 mm²

Gross area of connected leg $a_1 = 6 \times 70$
$$= 420 \text{ mm}^2$$

From equation (3.7) $a_2 = 819 - 420$
$$= 399 \text{ mm}^2$$

Noting from *Cl. 3.4.3* that $K_e = 1.1$,

$$A_e = 1.1 \times 819$$
$$= 901 \text{ mm}^2 \text{ but } \leq 819 \text{ mm}^2$$
$$\therefore \quad \text{use } A_e = 819 \text{ mm}^2$$

From equation (3.7) $P_T = 355 (819 - 0.3 \times 399)$
$$= \underline{248 \text{ kN}} \text{ which is satisfactory}$$

In this case the change from welded to a bolted end connection has necessitated an increase in section size amounting to some 30% by weight (using the 6 mm section as the basis). Although this means that a more expensive member is required, since steel prices for similar section types more or less follow the weight, the designer must base his choice of arrangement on the wider issues of the ease and practicality of both fabricating the member and making the connection on site. With the general tendency in the UK for the ratio labour costs/material costs to continually increase, the bolted arrangement will frequently prove to be the better, i.e. more cost-effective, choice.

Exercises

1   Select the lightest square hollow section from the *Structural Steel Handbook* in S355 steel capable of carrying a factored axial tensile load of 730 kN, assuming that full-strength welded end connections are provided.

[90 × 90 × 6.3 mm]

2   Determine the tensile capacity of an 80 × 80 × 8 mm angle section in S275 steel assuming that it contains a splice in which cover plates are provided to both legs. Assume the use of one row of M20 bolts in each leg arranged in pairs, i.e. not staggered.

[290 kN]

3   Determine the tensile capacity of the flat bar tie in the arrangement shown in Figure 3.8 assuming S275 steel and M20 bolts.

[1272 kN]

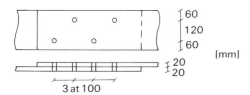

Figure 3.8

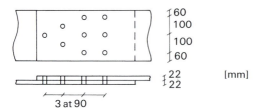

Figure 3.9

4   Determine the tensile capacity of the flat bar tie in the arrangement
    shown in Figure 3.9 assuming S355 steel and M20 bolts.

[2394 kN]

5   Show that a $100 \times 100 \times 12$ mm equal angle section in S275 steel is
    capable of carrying an axial tension of 450 kN assuming the use of a
    welded connection on one leg only.

[536 kN]

6   Determine the tensile capacity of a $150 \times 90 \times 10$ mm angle in S275
    steel assuming:
    (a) Connection through the longer leg by 2 rows of M20 bolts

[402 kN]

    (b) Connection through the shorter leg by one row of M24 bolts

[355 kN]

7   Determine the tensile capacity of a pair of $150 \times 75 \times 12$ mm angles
    having the long legs back to back, assuming that 2 rows of M20 bolts
    are used and that the steel is S275.

[908 kN]

References

1   Kitipornchai, S. and Woolcock, S.T. (1985) Design of Diagonal Roof Bracing
    Rods and Tubes, *Journal of the Structural Division ASCE*, III, No. 5, 1068–94.
2   CIDECT (1981) *The Strength and Behaviour of Statically Loaded Welded Con-
    nections in Structural Hollow Sections*, Monograph No. 6, Comité International
    pour le Développement et L'Étude de la Construction Tubulaire.
3   Kulak, G., Adams, P.F. and Gilmor, M.I. (1990) *Limit States Design in Struc-
    tural Steel*, 4th edn, Canadian Institute of Steel Construction.
4   Regan, P.E. and Salter, P.R. (1984) Tests on welded-angle tension members,
    Structural Engineer, **62B**(2), 25–30.
5   Nelson, H.M. (1953) *Angles in Tension*. Publication No. 7, British Construc-
    tional Steelwork Association, pp. 9–18.
6   Steel Construction Institute (1997) *Steelwork Design Guide to BS 5950: Part 1:
    1990. Volume 1 Section Properties, Member Capacities*, 4th edn.

Chapter 4

Axially loaded columns

One of the most frequently encountered and basic types of structural member is the column whose main function is the transfer of load by means of compressive action. Two common examples drawn from the wide range of structures in which such members are found are shown in Figure 4.1. Depending upon the precise way in which the column is joined to the neighbouring parts of the structure, it may also be required to carry bending moments. Nevertheless, a proper appreciation of the behaviour of members in pure compression forms an important first step in understanding this more general problem because design for combined loading (considered in Chapter 6), i.e. compression and bending, is usually based upon considerations of the interaction of the various individual load components.

The response of a compression member to a nominally axially applied load depends upon a number of factors, the most important of which are its length and cross-sectional shape, the characteristics of the material from which it is made, the conditions of support provided at its ends and the method used for its manufacture. Table 4.1 lists the major forms of response.

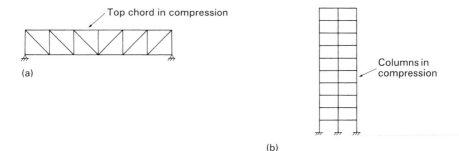

Figure 4.1 Examples of compression members: (a) compression members in a truss; (b) compression members in a building frame.

Table 4.1 Possible failure modes for an axially loaded column

Mode	Description	Illustration	Section	Comments
Squashing	Providing the length is relatively small (stocky column) and its plate elements are not too thin (compact cross-section) then the column will be capable of attaining its squash load (yield stress × area)		4.1	Member needs to be extremely stocky
Overall flexural buckling	Failure occurs by excessive deflection in the plane of the weaker principal axis, the load at which this occurs becoming progressively less as the column slenderness is increased		4.2	Controls the design of most compression members
Torsional buckling	Failure occurs by twisting about the longitudinal axis			This mode is unlikely for hot-rolled sections or for fabricated sections of 'normal' proportions but may be important for lighter cold-formed members, particularly unsymmetrical shapes
Local buckling	Failure occurs by buckling of one or more individual plate elements, e.g. flange or web, with no overall deflection; this may be prevented by placing suitable limits on plate width-to-thickness ratios; alternatively, where such limits are exceeded, the design strength must be reduced		4.1.2	Proportions of normal hot-rolled sections are such as to preclude this in most instances; needs to be considered for fabricated members or cold-formed sections
Local failure	Where compound members are formed by joining together two or more shapes to form a lattice cross-section, failure of a component member may occur if the joints between members are too widely spread		4.4	Design rules usually require intermediate fastening to be sufficient to permit design for overall buckling as an equivalent solid strut

4.1   Stocky columns

4.1.1   Stub column behaviour

The results of a typical laboratory compression test on a short length of rolled section are shown in Figure 4.2 in the form of a load versus end-shortening curve. Such a test is often referred to as a 'stub column test'. Comparison with the results of a compression test on a small coupon cut from the cross-section (presented previously as Figure 1.6) shows that the major difference between the two is the lower limit of proportionality exhibited by the test on the full cross-section. The explanation for this lies in the non-uniform yielding of the stub column caused by the presence of residual stresses [1]. Thus those fibres which contain residual compression have their effective yield point reduced, while those containing residual tension have theirs increased. In both cases, however, the full strength of the material can be achieved with the stub column failing at its squash load. Although the actual collapse of the stub column would normally be precipitated by local buckling as illustrated in Figure 4.3, for compact sections this would not occur until after considerable plastic straining had taken place. Many thousands of stub column tests have now been conducted and these demonstrate quite conclusively that the appropriate basis for the design of stocky columns of compact cross-section is the squash load.

4.1.2   Local buckling in columns

Not all stocky columns will be capable of attaining their full squash load. If the individual plate elements which make up the cross-section, for example the web and the two flanges in the case of an I-section, are thin,

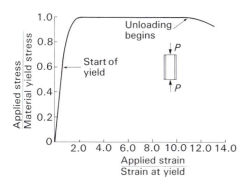

Figure 4.2   Stress–strain behaviours of a full section in compression – stub-column response.

Figure 4.3

Exposed columns support the roof of Princes' Square.

then local buckling of the type shown in Figure 4.4 may occur at a lower load. Analysis of this type of failure is somewhat complex so design rules are based largely on experimental data. For columns (which is the only case dealt with here – see Chapter 5 for more details) it is frequently possible to simply 'design out' the problem by so limiting the proportions of the component plates that local buckling effects will not influence the cross-section's strength. In cases where more slender plating is to be used the section's strength must be suitably reduced.

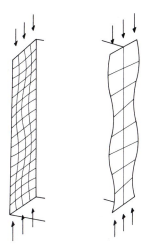

Figure 4.4   Local buckling in box and I-section columns (deformations of a single flange only shown). (*BSC Teaching Project, Imperial College, 1985.*)

(a) Design approach
Clause 3.5.2 of BS 5950: Part 1 classifies those sections for which yield may be attained without prior local buckling as semi-compact. Upper limits for this range are:

Flanges, i.e. plate elements   Webs, i.e. plate elements
supported along one   supported along both
longitudinal edge   longitudinal edges
$b/T \not> 15\sqrt{(275/p_y)}$ $b/T \not> 40\sqrt{(275/p_y)}$ for a rolled section
$b/T \not> 13\sqrt{(275/p_y)}$ $b/T \not> 40\sqrt{(275/p_y)}$ for a welded section

where the method to be used to assess b, d and T is given in *Figure 5*. Stricter limits are imposed for welded flanges in recognition of the weakening effect of the more severe residual stress present [2]. Sections which

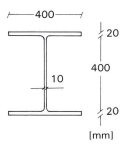

Figure 4.5   Column cross-section of Example 4.2.

do not meet these limits are classified as 'slender' and assessment of their load-carrying capacity must reflect the influence of local buckling. *Clause 3.6* explains how this is treated in design by working with an effective cross-section (of reduced area) when calculating section properties. *Figure 8* defines this effective section for the most common structural shapes. Particular care should be taken on those rare occasions when the slender cross-section contains one or more slender outstand elements. However, relatively few rolled sections are affected when using other than the higher grades of steel. In particular, no UC sections in S275 steel are less than semi-compact.

Example 4.1

A 305 × 102 mm UB33 is to be used as a short column carrying axial load. Is its compressive strength likely to be affected by local buckling assuming (a) S275 steel, (b) S355 steel?

Solution
From section tables, B = 102.4 mm, T = 10.8 mm, d = (275.8 − 2 × 7.6) mm, t = 6.6 mm, A = 40.8 cm².

Reference to *Figure 5* shows that T, d and t correspond to those used in *Table 11*.

$$b = \tfrac{1}{2}(102.4 - 6.6 - 2 \times 6.6) = 41.3 \text{ mm}$$
$$b/T = 41.3/10.8 = 3.82$$

From *Table 11*, for p_y = 275 N/mm² limit is 15

$$d/t = 260.6/6.6 = 39.5$$

From *Table 11*, for p_y = 275 N/mm² limit is 40
   Full cross-section is available and P_c = 275 × 4080 N
$$= 1122 \text{ kN}$$
From *Table 11* for p_y = 355 N/mm², flange limit is $15\sqrt{(275/355)}$ = 13.2
From *Table 11* for p_y = 355 N/mm², web limit is $40\sqrt{(275/355)}$ = 35.2
Therefore cross-section is slender on account of proportions of the web *Cl. 3.6.2* provides a method to allow for this.
   By inspection section contains no internal element wider than 80 εt and no class 4 outstand elements. Effective cross-section is therefore as indicated by *Figure 8a*.

$$A_{eff} = 4080 - (260.6 - 40 \times 6.6 \times \sqrt{275/355})6.6$$
$$= 4080 - (260.6 - 232.4)6.6$$
$$= 4080 - 186.1 = \underline{3894 \text{ mm}^2}$$
$$P_c = 355 \times 3894 = \underline{1382 \text{ kN}}$$

In this case local buckling reduces the compression strength by about 5%. Since the web proportions control and most of the section's area is concentrated in the (semi-compact) flanges, allowance for the reduced effectiveness of the web leads to a much smaller loss of design capacity.

Example 4.2

Check whether the welded column section shown in Figure 4.5 could be designed for its full squash load. Assume S355 steel.

Solution
From *Table 11* the flange limit is $13\sqrt{(275/345)} = 11.7$
Actual b/T, noting how b is defined in *Figure 5*, $= (200 - 12.5)/20 = 9.38$
Web limit from *Table 11* is $40\sqrt{(275/345)} = 35.7$
Actual d/t, noting how d is defined in *Figure 3*, $= 400/10 = 40$
Web is slender. Therefore either reduce design resistance by using an effective cross-section or replace 10 mm web by one of at least $(400/35.7) = 11.2$ mm; use 12 mm web and design for full squash load.

4.2   Slender columns

4.2.1   Background to the problem

Although the theory of the elastic stability of perfect pin-ended struts [3, 4], sometimes referred to as the Euler theory, provides some insight into the behaviour of slender compression members, it omits the consideration of a number of important factors [5]. These are often grouped under the general heading of 'imperfections' and include such factors as initial lack of straightness, accidental eccentricities of loading, residual stresses and variation of material properties over the cross-section. Their combined effect is to produce the type of relationship between theory and experiment shown in Figure 4.6. Thus, while very slender columns fail at loads which are close to their elastic critical load, columns of intermediate slenderness (which account for a large proportion of cases found in actual construction) collapse at loads some way below either the elastic critical load or the squash load. Only by resorting to complex numerical methods is it possible for an analysis to include the effects of these imperfections.

The design, as opposed to the analysis, of columns is usually based on the concept of one or more 'column curves' which give load-carrying capacity directly as a function of slenderness. Figure 4.7 presents the set of four curves in BS 5950: Part 1. These have been based on the careful study [7–10] of both theoretical and experimental data. The reason for using

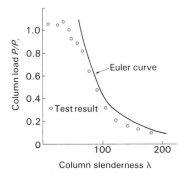

Figure 4.6   Typical column test data compared with basic Euler strut theory, data on high-strength H-sections, from reference [6].

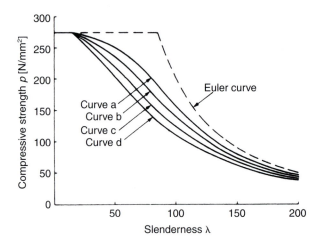

Figure 4.7   Column design curves of BS 5950: Part I, $p_y = 275$ N/mm².

more than one curve becomes clear when data of the form shown in Figure 4.8 are examined. Despite the inevitable scatter associated with column tests, the results indicate clear differences in strength between columns of the same slenderness but different type. This is largely a consequence of the different ways in which progressive yielding affects the stiffness of the various shapes; a factor that is itself dependent upon the pattern of residual stresses present. This in turn depends upon the method of manufacture which will also influence other controlling factors such as straightness and dimensional tolerances. Thus, in common with other modern national codes, BS 5950: Part 1 recognizes this fact by requiring the use of different column curves for different classes of column.

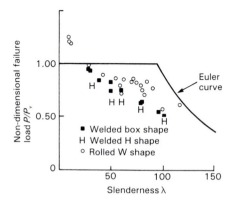

Figure 4.8 Experimental data for the column strength of different types of steel section. (*Chen and Atsuta, Theory of Beam Columns, vol. I. McGraw-Hill 1976* [11], by permission.)

4.2.2 Design approach

A formula describing the four curves of Figure 4.7 is presented in *Appendix C* of BS 5950: Part 1. However, it is not necessary to use this in actual design (unless column design is to be programmed) since tables of design axial strength p_c versus slenderness $\lambda = l/r_{min}$, in which r_{min} is the minimum radius of gyration, for each curve are given for a complete range of yield strengths, p_y in *Table 24*. The particular table, i.e. whichever column curve should actually be used, must first be ascertained by reference to the selection table, *Table 23*. The following worked examples illustrate the process.

Example 4.3

Calculate the compressive resistance of a 203 × 203 mm UC60 of height 3.1 m. Assume that the conditions at both ends of the *xx* and *yy* planes are such as to provide 'simple support'. Take the design strength of the steel p_y as 275 N/mm².

Solution
Unless the axis about which buckling will occur is obvious all possibilities must be checked. For UC sections r_y is normally between about one third and one half of r_x so that the likely mode of failure is by buckling about the minor axis. However, in cases where different effective lengths apply for the two planes both possibilities should normally be checked.

From section tables, $A = 75.8$ cm², $r_y = 5.19$ cm, $r_x = 8.98$ cm. Work in mm and N.

$$\lambda = l/r_y = 3100/51.9 = 59.7$$

From *Table 23* for a UC buckling about the minor axis, curve *c* is appropriate. Therefore from *Table 24* for $\lambda = 59.7$ the corresponding value of the axial strength p_c is 201 N/mm².

Hence compressive resistance $P_c = 201 \times 7580 = 1524 \times 10^3$ N
$$= \underline{1524 \text{ kN}}$$

Clearly for this example there is no real need to check for buckling about the major axis since $r_x > r_y$. It is left to the reader to show that this is indeed the case by using column curve *b* to find that $P_c = 1948$ kN.

Example 4.4

Repeat the previous example for the 200 × 200 × 59.9 mm equal leg angle section shown in Figure 4.9.

Solution
Since the principal axes for an angle section do not coincide with the rectangular $x - x$ and $y - y$ axes the buckling strength about the minor principal axis $v - v$ should normally be checked.
From *Table 23*, curve *c* is appropriate for buckling about any axis.

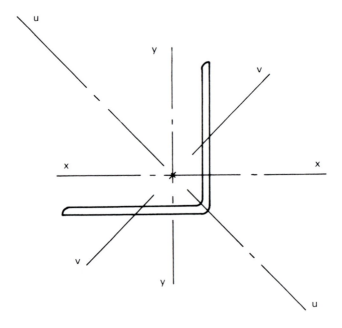

Figure 4.9   Equal leg angle section of Example 4.4.

Therefore only the axis about which the slenderness is greatest need be considered.

From section tables, $r_{xx} = r_{yy} = 6.11$ cm, $r_{vv} = 7.79$ cm

Work in mm and N.

Max. $\lambda = 3100/39.2 = 79.1$

From *Table 24* for $\lambda = 79.1$ and $p_y = 275$ N/mm^2.

$p_c = 163$ N/mm^2.

Hence $P_c = 163 \times 7630 = 1244 \times 10^3$ N = 1244 kN

This is approximately 17% less than the strength of the UC section of almost identical weight. This is a direct result of the less favourable arrangement of material with regard to bending stiffness leading to a lower value for r_{min}. However, as explained in Section 4.4 angles are frequently used as compression members in lightly loaded trusses because of the relative ease of making connections between them.

4.2.3 Welded sections

The column curves of Figure 4.7 are intended for application to hot-rolled shapes. Available data for welded shapes [5, 10, 12] show that because of the rather different pattern of residual stresses present (see Figure 4.10), a column curve of a slightly different shape should be used. Rather than increase the number of column curves still further, *Cl. 4.7.5* of BS 5950: Part 1 deals with this problem by the simple expedient of requiring welded columns to be designed as if their yield strength were $p_y - 20$ N/mm^2. This device leads to the correct sort of design strengths over much of the range [7] although it does, of course, produce an inconsistency for very stocky columns which cannot be designed for their full squash load.

In cases where I- or H-sections are welded together from flame-cut plates the effect of the flame-cutting will be to produce beneficial tensile residual stresses at the flange tips as shown in Figure 4.10(c). *Tables 23* and *24* therefore permit design to be based on the full value of p_y in such cases.

Example 4.5

A heavy column is required to support a gantry girder and a special H-section is to be fabricated. The trial section is shown in Figure 4.11. Check its suitability to support a factored axial load of 32 000 kN assuming both ends to be pinned over a length of 8 m. Steel of design strength 325 N/mm^2 is to be used. Could a rolled section be suitably reinforced (by welding cover plates to its flanges) so as to provide an alternative section?

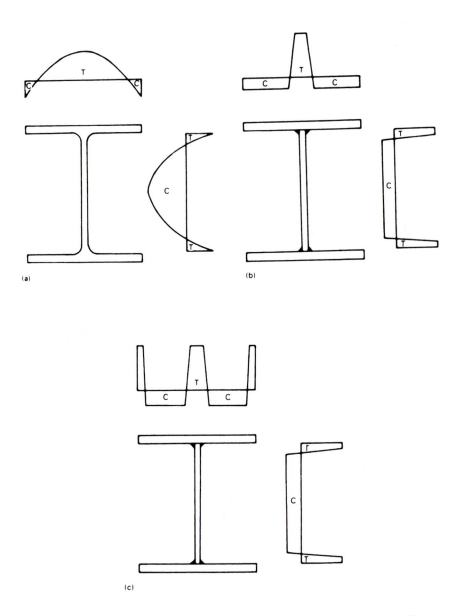

Figure 4.10 Typical residual stress patterns in column sections made by different processes: (a) rolled section; (b) welded section; (c) welded section using plates with flame cut edges.

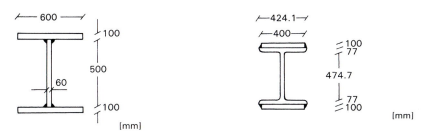

Figure 4.11   Column cross-section of Example 4.5, welded section.

Figure 4.12   Column cross-section of Example 4.5, reinforced rolled section.

Solution

$A = (60 \times 10) \times 2 + 50 \times 6 = 1500 \text{ cm}^2$

$I_y = 2(10 \times 60^3)/12 = 360 \times 10^3 \text{ cm}^4$ (neglects web)

$r_y = \sqrt{(360 \times 10^3/1500)} = 15.49 \text{ cm}$

$\lambda = L/r_y = 8000/155 = 51.6.$

From *Table 23*, noting that $t > 40$ mm, use curve *d*.
Since the section will be fabricated by welding and no guarantee that flame-cut plates are to be used is provided, use a reduced design strength

$p_y = 325 - 20 = 305 \text{ N/mm}^2.$

Table 24 for $\lambda = 51.6$ and $p_y = 305 \text{ N/mm}^2$ gives $p_c = 217 \text{ N/mm}^2$
$P_c = 217 \times 1500 \times 10^2 = 32\ 550 \times 10^3 \text{ N} = \underline{32\ 500 \text{ kN}}$, section is suitable.
The heaviest rolled section is a 356 × 406 UC 634, the relevant properties for which are $A = 808 \text{ cm}^2$, $I_y = 98\ 211 \text{ cm}^4$ and $r_y = 11.0 \text{ cm}$. Since this provides about one half of the area of the welded section it will need substantial cover plates. As a first trial use 400 mm × 100 mm plates on both flanges as shown in Figure 4.12.

$A = 808 + 2(10 \times 40) = 1608 \text{ cm}^2$

$I_y = 98\ 211 + 2(10 \times 40^3)/12 = 204\ 878 \text{ cm}^4$

$r_y = 204\ 878/1608 = 11.29 \text{ cm}$

$\lambda = L/r_y = 70.9$

From *Table 23*, noting section is as shown in *Figure 14*, and $t > 40$ mm use curve *b*. The full p_y may be used.
From *Table 24*, for $\lambda = 70.9$ and $p_y = 340 \text{ N/mm}^2$, $p_c = 233 \text{ N/mm}^2$
Hence compressive resistance $P_c = 233 \times 160\ 800 = 37\ 466 \times 10^3 \text{ N}$
$$= \underline{37\ 466 \text{ kN}}$$

Since this exceeds the required resistance the section could be redesigned using smaller cover plates. It is left to the reader to show that

370×100 mm plates provide a compressive resistance of 34 830 kN for an area of 1548 cm^2 and a slenderness of 73.7.

This example clearly demonstrates the effect of making allowance for the variations in strength between different types of column. Although the areas of the two sections are similar, the reinforced UC is significantly more slender (λ of 70.9 compared with 51.6) and yet the compressive strengths of the two sections are very similar (233 N/mm^2 and 217 N/mm^2). The reason for this lies in the more favourable column curve assigned to the cover plated section as well as the use of a reduced design strength for the welded section.

The choice of section for a given application will depend on a number of factors, especially availability of materials and fabrication facilities, although it is worth noting that the reinforced section would occupy less space on plan.

4.3   Influence of end conditions

In discussing the column curves of the previous section it was assumed throughout that both ends were supported such that:

1   they could not translate with respect to one another
2   no rotational restraint was present.

While conditions in practice may sometimes approximate to this, several other arrangements will also be encountered. True ultimate strength results for columns with other than pinned ends are not readily available. Even if they were it would still be necessary to devise a simplified treatment for use in design since the provision of a portfolio of column curves to cover all possible restraint conditions would be impractical. The usual approach for design consists of reducing the actual case under consideration to an equivalent pin-ended case by means of an effective-length factor determined from a comparison of elastic critical loads. This process therefore assumes that the influence of imperfections will be broadly similar for all forms of restraint, being a function of effective slenderness only.

The notion of an effective column length comes directly from elastic stability theory [3] where it is used as a device to relate the behaviour of columns provided with any form of support to the behaviour of the basic pin-ended case. Thus the general expression for critical load becomes

$$P_{cr} = \pi^2 EI/l^2 \tag{4.1}$$

where $l = kL$ is the effective length and k is termed the 'effective length factor'.

Taking, as an example, the case of a column with fixed ends, for which the critical load is

$$P_{cr} = 4\pi^2 EI/L^2 \tag{4.2}$$

the effective length is obtained directly as

$$l = L/2, \text{ or } k = 0.5$$

Table 4.2 gives theoretical effective length factors for several standard cases. When used in the context of elastic critical loads the effective length also corresponds to the distance between points of inflection in the buckling mode [4]. An important general point to note from Table 4.2 is that when relative translation of the ends is prevented, k cannot exceed unity but that effective lengths up to several times the actual column height are possible for columns which are free to sway.

4.3.1   Design approach

Guidance on the choice of effective length factors for columns in simple construction is given in *Cl. 4.7.3* of BS 5950: Part 1, particularly *Table 22*. Comparison with Table 4.2 shows the code values to be either equal to or slightly higher than the equivalent theoretical values. When higher values are specified it is usually in recognition of the practical difficulties of providing complete restraint against rotation. Further information on the appropriate effective column lengths to use in single storey and multi-storey buildings of simple construction is provided in *Appendix D*. The decision as to what value is applicable to a particular case often requires

Table 4.2   Effective length factor for columns

	Both ends pinned	Intermediate restraint	One end fixed	Both ends fixed	Cantilever
Support arrangements					
Value of k based on elastic critical load	1.0	0.5	0.7	0.5	2.0
BS 5950: Part I design value of k	1.0	0.5	0.85	0.7	2.0

considerable judgement; for situations in which the designer is uncertain of the degree of restraint present, the safe approach is always to neglect the restraint and to select a high value for k.

Example 4.6

Repeat Example 4.3 assuming that the column is built in at its base and is supported at its top in such a way that deflection about the minor axis is prevented and deflection about the major axis is not.

Solution
Reference to *Table 22* shows that the appropriate effective lengths are

minor axis, $l_y = 0.85\ L$
major axis, $l_x = 2.0\ L$

Referring back to Example 4.3, $\lambda_y = 0.85 \times 3100/51.9 = 50.8$
$\lambda_x = 2.0 \times 3100/89.8 = 69$

Thus, because of the different degrees of restraint in the two planes, major-axis buckling is now more critical. From *Table 23* use curve b, hence using *Table 24* for $p_y = 275\ \text{N/mm}^2$ and $\lambda = 69$, value of $p_c = 204\ \text{N/mm}^2$ and $P_c = 204 \times 7580\ \text{N} = \underline{1546\ \text{kN}}$

Had the column also been restrained at its top about the major axis, then $\lambda_x = 29.3$ and minor-axis buckling would again have controlled, leading to $P_c = 1652$ kN. Comparing this with Example 4.3 shows that the change in restraint conditions (provision of rotational restraint at the base) produces an increase in strength of about 8%, at least the equivalent of a change in the column curve used.

4.4   Special types of strut

4.4.1   Angle sections

Single angles are often used as compression members in situations where comparatively low forces need to be transmitted, a common example being the roof truss shown earlier in Figure 4.1a. In situations where a single angle could not provide sufficient compressive strength, or perhaps where the disparity in size between tension and compression members would make jointing difficult, double angles may be used. These are formed from two angles placed back to back, normally with a space between them to allow the joints at either end to be made via gusset plates in such a way that eccentricity of loading at the joint is minimized. It is, of course, necessary to ensure that the two sections function together as one compound member. Thus 'stitching' must be provided at sufficient intermediate points that the load for buckling of one angle between fasteners exceeds the load for overall buckling of the compound section. In this, as in any problem involving buckling of a single angle, it is important to remember that the weakest plane will be in the direction of the minor principal axis which does not of course, coincide with either rectangular axis.

(a) Design approach

Rules for the design of angle struts are based largely on empirical data [9, 13, 14] due to the difficulties associated with quantifying both end restraint conditions and the eccentricities of loading introduced by the joints. For continuous struts, i.e. where one length is 'run through' to form several members as might happen for example in the rafter of a roof truss, it is permissible to design the intermediate bays as axially loaded, with the effective length being taken as the actual length in that bay. For discontinuous struts (including the end bays of continuous struts) BS 5950: Part 1 gives two procedures depending upon the type of end fixing. In the case of single angle struts (*Cl. 4.7.10.2*) these are:

1   connection through one leg by two or more fasteners in line or the equivalent in welding,

$$\lambda = 0.7 \, L/r_{aa} + 30 \not< 0.85 \, L/r_{vv}$$

in which r_{aa} = radius of gyration about an axis through the centroid of the angle parallel to the gusset,

r_{vv} = the minimum radius of gyration

2   single fastener or the equivalent in welding,

$$\lambda = 0.7 \, L/r_{aa} + 30 \not< 1.0 \, L/r_{vv}$$

and in addition, $P_c \not> 0.8 \, p_c A$.

Whereas the first of these includes some allowance for eccentricity of loading by using a pessimistic effective length, the second, because it would clearly be confusing to specify an effective length factor greater than 1.0 when end translation is prevented, allows for the (probably greater) effect of load eccentricity by assuming that part of the compressive resistance must be used to resist bending, a device that is similar to the use of effective area for tension members as described in Chapter 3. Similar rules are also given in *Cl. 4.7.10.3* for double angle struts. Because of the smaller eccentricities associated with this class of section these are less severe.

Example 4.7

Determine the compressive resistance of an $80 \times 80 \times 10$ equal-angle section in S275 steel when it is used as a strut over a length of 1.8 m. Assume a single fastener is provided at each end.

Solution
From section tables, $A = 15.1$ cm^2, $r_{min} = r_{vv} = 1.55$ cm
From *Cl. 4.7.10.2*, take $\lambda = 1.0 \times 1800/15.5 = 116$
From *Table 23*, use curve c
From *Table 24*, for $\lambda = 116$ and $p_y = 275$ N/mm^2, value of $p_c = 102$ N/mm^2
From *Cl. 4.7.10.2*, $P_c = 0.8 \times 102 \times 1510$ N $= \underline{123\ \text{kN}}$

If the end connections had been improved to two fasteners in line then P_c could be increased to 192 kN, an improvement of over 50%.

A summary of appropriate values of λ for angle, channel and tee-struts with various forms of eccentric connection, is provided in *Table 25*.

4.4.2   Laced and battened struts

The columns of industrial buildings are often called upon to provide support for a gantry crane. Quite heavy axial loads are therefore introduced into the lower portions of these columns. Rather than use a heavy section over the full height, a second member may be introduced over this lower length and the two legs connected together into the lattice arrangement shown as Figure 4.13. Two slightly different forms may be used:

1   the laced column in which relatively light transverse members are arranged in a triangulated fashion;
2   the battened column in which rather heavier battens are placed only at right angles to the column axis.

(a)   Design approach
Design of laced and battened struts is similar in principle to the design of double angle struts in that the lacing or battens should be so arranged that they insure against premature local failure [15]. The strut may then be designed as a single integral member with a slenderness given by

$$\lambda = \sqrt{(\lambda_m^2 + \lambda_c^2)}$$

where $\lambda_m = l/r$ for the whole member
        $\lambda_c = l/r_{min}$ for the main component
subject to the limitations $\lambda \not> 50$ and $\lambda \not< 1.4\,\lambda_c$.

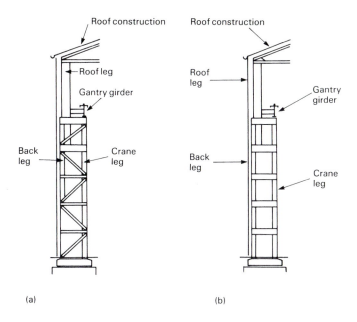

Figure 4.13 Compound columns suitable for supporting crane gantries in industrial buildings: (a) laced column, (b) battened column. (*Bates, Constrado Publications.*)

Additional rules covering the proportioning of the transverse members and the arrangement of the fasteners are given in *Cl. 4.7.8, 4.7.9* and guidance on the assessment of effective lengths for intermediate portions of the main legs is given in *Appendix D*.

Exercises

1   Check whether a 406 × 140 UB 39 in S275 steel would be affected by local buckling effects when used as a column.

[web limit exceeded, $P_c = 1107$ kN]

2   Determine the axial load capacity of a short length of square box column in S355 steel fabricated by welding together four 800 × 20 mm plates.

[2272 kN]

3   Determine the capacity of a 254 × 254 UC 107 in S275 when used as an axially loaded column of effective length 4.2 m.

[2636 kN]

4   Select the lightest UC in S275 steel that is capable of carrying an axial compressive load of 2100 kN.

[305 × 305 UC 118]

5   Determine the axial load capacity of a 90 × 90 × 8 mm angle section in S275 steel when used as a column with an effective length of 1.2 m.

[257 kN]

6   Select the lightest equal leg angle in S275 steel capable of carrying an axial compressive load of 295 kN over an effective height of 1.25 m.

[100 × 100 × 8 mm]

7   Determine the load-carrying capacity of a box section made from four 800 × 20 mm S275 steel plates when used as an axially loaded column over an effective height of 10 m.

[10 656 kN]

8   Determine the compressive resistance of a 120 × 120 × 10 mm angle of S275 steel when used as a strut over a length of 2.2 m, assuming:
(a) Fastening with a single bolt at each end

[318 kN]

(b) Fastening with two bolts in line at each end

[378 kN]

9   Select an unequal angle section in S275 steel capable of sustaining an axial compressive load of 255 kN over a length of 1.8 m, assuming:
(a) Fastening to a gusset through the longer leg with a single bolt

[150 × 90 × 10 mm]

(b) Fastening to a gusset through the longer leg with at least two bolts in line.

[125 × 75 × 10 mm]

References

1 Structural Stability Research Council (1999) Technical Memorandum No. 3: Stub-column test procedure. Appendix B in T.V. Galambos (ed.) *Guide to Stability Design Criteria for Metal Structures*, 5th edn, Wiley-Interscience, New York.

2 Dwight, J.B. and Moxham, K.E. (1969) Welded steel plates in compression, *Structural Engineer*, **47**(2), 49–66.

3 Timoshenko, S.P. and Gere, J.M. (1961) *Theory of Elastic Stability*, 2nd edn, McGraw-Hill, New York.

4 Kirby, P.A. and Nethercot, D.A. (1979) *Design for Structural Stability*, Granada, St Albans.

5. Tall, L. (1982) Centrally compressed members, in R. Narayanan (ed.) *Axially Compressed Structures – Stability and Strength*, Applied Science Publishers, London, pp. 1–40.

6 Strymowicz, G. and Horsley, P. (1969) Strut behaviour of a new high yield stress structural steel, *Structural Engineer*, **47**(2), 73–8.

7 Dwight, J.B. (1978) Strength in Compression, Revision of BS 449, *The Structural Use of Steelwork in Building Symposium*, Institution of Structural Engineers, London, pp. 11–16.

8 Dwight, J.B. (1975) Adaptation of Perry formula to represent the new European steel column-curves, *Steel Construction, AISC*, **9**(1).

9 ECCS (1977) Manual on the Stability of Steel Structures, Introductory Report, *Second International Colloquium on Stability*, Liège.

10 Galambos, T.V. (ed.) (1999) *Guide to Stability Design Criteria for Metal Structures*, 5th edn, Wiley-Interscience, New York.
11 Chen, W.F. and Atsuta, T. (1976) *Theory of Beam-Columns*, Vol. 1, McGraw-Hill, New York.
12 Young, B.W. and Robinson, K.W. (1975) Buckling of axially loaded welded steel columns, *Structural Engineer*, **53**(5), 203–7.
13 Kennedy, J.B. and Madugula, M.K.S. (1982) Buckling of single and compound angles, in R. Narayanan (ed.) *Axially Compressed Structures – Stability and Strength*, Applied Science Publishers, London, pp. 181–216.
14. Woolcock, S. and Kitipornchai, S. (1980) The design of single angle struts, *Steel Construction, AISC*, **14**(4), 2–23.
15 Porter, D. (1982) Battened columns – recent developments, in R. Narayanan (ed.) *Axially Compressed Structures – Stability and Strength*, Applied Science Publishers, London, pp. 249–78.

Chapter 5

Beams

One of the most frequently encountered types of structural member is the beam, the main function of which is to transfer load principally by means of flexural or bending action. In a typical rectangular building frame the beams would comprise the horizontal members which span between adjacent columns; secondary beams might also be used to transmit the floor loading into the main beams. For the more usual forms of structural framing it is normally sufficient to consider only bending effects, the influence of any torsional loading on the beams being relatively slight. Certain types of problem, such as design of crane girders do, however, require a proper allowance to be made for the effects of torsion.

For guidance on problems combining bending and torsion, including ways of minimizing unwanted torsional effects by appropriate detailing of the load transfer into the beam, reference should be made to the appropriate SCI design guide [1].

The main forms of response for a beam subjected to simple uniaxial bending are listed in Table 5.1. Which of these will govern in a particular case depends principally upon the proportions of the beam, the form of the applied loading and the type of support provided. In addition to satisfying these strength limits it is also necessary to ensure that the beam does not deflect too much under the working loads, i.e. to satisfy the serviceability limit state.

5.1 In-plane bending of beams of compact cross-section

This discussion assumes that the beam's cross-section is such that the effects of local buckling may be neglected (a full discussion of this topic is presented in Section 5.3.1). The behaviour of a simple beam, which is constrained to deflect in the plane of the applied loading under the action of a gradually increasing bending moment, is illustrated in Figure 5.1. Neglecting, for the present, the effect of residual stresses, the beam's response will be linear up to that value of the applied load W_y which just causes the

Table 5.1 Main failure modes for beams

Mode	Description	Illustration	Section	Comments
Excessive bending	Providing the beam is adequately braced in the lateral plane (stocky beam) and its component plate elements are not too thin (compact cross-section), then failure will take place by excessive deformation in the plane of the applied loading		5.1	Basic mode of failure if all others are prevented
Lateral torsional buckling	Failure occurs by a combination of lateral deflection and twist, the load at which this occurs being dependent upon the proportions of the beam, the way the loading is applied and the support conditions provided		5.2	Can be prevented by the provision of suitable lateral bracing
Local buckling	Failure occurs by buckling of a flange on compression or of the web due to shear or combined shear and bending or, where concentrated loads are applied, as a result of vertical compression		5.3 / 5.3.1	Unlikely for hot-rolled sections for which the proportions have been selected so as to minimize the importance of flange and web buckling; web stiffening sometimes required to prevent shear buckling in plate girders, bearing stiffeners sometimes required under point loads and at reaction points
			5.3	Likely only for short spans and/or deep beams; can be prevented by suitable web stiffening.
Local failure	Several possibilities including: (i) shear yield of web (ii) local crushing of web (iii) excessive curling of thin flanges (iv) local failure around web openings (if present). Special provision may be required around large web holes, e.g. use of local reinforcement		5.4	Possible only for extreme sections with very wide flanges.

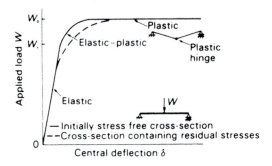

Figure 5.1   Behaviour of simply supported steel beam. (*BSC Teaching Project, Imperial College, 1985.*)

maximum extreme fibre stress at the cross-section of greatest moment to reach the material yield strain ϵ_y. At higher loads, deformations will increase more rapidly until the fully plastic moment M_p is reached at the most highly stressed cross-section, whereupon a plastic hinge will form under the load. According to simple plastic theory [2, 3] deformations will now become uncontrolled. In practice the load-carrying capacity may actually be slightly greater due to the effects of strain hardening. However, it is customary to neglect this in design so that for simple beams, namely those that are not supported in such a way that redistribution of moment may occur (see Chapter 11) the formation of a plastic hinge at one point corresponds to the attainment of the ultimate load. The effect of the residual stresses which are normally present in structural sections is to cause yielding to start at a lower load with a consequent increase in the deflections which occur at all subsequent load levels. However, the value of W_p is not affected because the residual strains must themselves be in equilibrium and cannot therefore alter the value of M_p.

Since the design for bending of laterally braced beams, i.e. those for which failure is governed by plastic action, is treated in BS 5950: Part 1 as a special case of the more general problem involving consideration of lateral–torsional buckling, comments on the design approach will be delayed until Section 5.2.

5.2   Lateral–torsional buckling of beams of compact cross-section

5.2.1   Background to the problem

In much the same way that the design of all but the most stocky struts is controlled largely by considerations of overall instability, so the design of most beams must be undertaken with a view to ensuring an adequate degree of safety against overall buckling. For beams the form of instability is, however, rather more complex since it involves both lateral deflection and twist as shown in Figure 5.2. For the ideal case of a perfectly straight beam, loaded exactly in the plane of the web, theory [4–7] tells us that at the elastic critical load the beam will fail suddenly by deflecting sideways and twisting about its longitudinal axis – a form of response that may be observed in laboratory tests. Although the basic theory provides an adequate description of the behaviour of beams tested under very carefully controlled laboratory conditions, it does not cater for several of the factors which affect the lateral stability of beams in actual structures. Among the more important of these are initial bow and initial twist in the section, accidental eccentricities of loading and premature yielding due to the presence of residual stresses. Therefore, whilst elastic buckling theory assists in the identification of the governing parameters of the problem, proper use must also be made of representative test data if satisfactory design rules are to be established.

Experiments have demonstrated clearly that beams with closely spaced restraints can reach M_p while long unrestrained spans effectively fail by elastic lateral–torsional instability at moments that are very close to M_E, the theoretical elastic critical value [4–7]. Using the ratio of these two

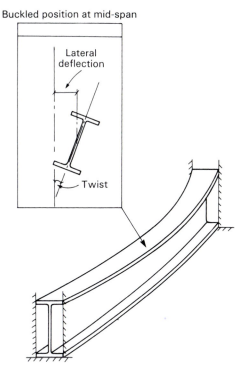

Figure 5.2   Lateral–torsional buckling of a beam.

quantities as a measure of a beam's proneness to lateral–torsional collapse leads to the pictorial display of the problem shown in Figure 5.3, where the quantity $(M_p/M_E)^{\frac{1}{2}}$ may be regarded as an 'effective slenderness for lateral–torsional buckling'. When test data are plotted on this basis it becomes possible to distinguish three regions of beam behaviour.

1   Stocky beams: $(M_p/M_E)^{\frac{1}{2}} < 0.4$ for which M_p may be attained. (Beams for which plastic hinge action is possible are a subset of this requiring more closely specified limits; this topic is discussed in Chapter 11.)
2   Beams of intermediate slenderness: $0.4 < (M_p/M_E)^{\frac{1}{2}} < 1.2$ which collapse through the combined effects of plasticity and instability at moments below either M_p or M_E.
3   Slender beams: $(M_p/M_E)^{\frac{1}{2}} > 1.2$ which buckle at moments approaching M_E.

In the foregoing explanation it has simply been assumed that M_E corresponds to the theoretical elastic critical moment for the particular beam under consideration. Examination of the background theory [4–7] tells us that this quantity is a complex function of a number of parameters, the most important of which are the beam geometry, in particular its bending and torsional properties and its span, the type of restraint provided in the lateral plane and the pattern of moments (which will, of course, be affected by the conditions of support provided in the transverse plane). Thus the type of presentation of lateral buckling data used in Figure 5.3 enables all of these factors to be conveniently accounted for.

5.2.2   Design approach

The basic design condition to ensure sufficient strength against overall buckling is given in *Cl. 4.3.6.2* of BS 5950: Part 1 as

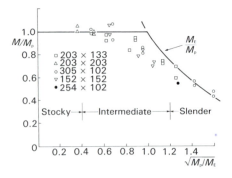

Figure 5.3   Lateral–torsional buckling strength of steel beams of Gr. 55 steel. (*BSC Teaching Project, Imperial College, 1985.*)

$$M_x \leq M_b/m_{LT} \tag{5.1}$$

in which M_b is the lateral-torsional buckling resistance moment
M_x is the maximum moment about the major axis
m_{LT} is the equivalent uniform moment factor for lateral-torsional buckling

In addition M_x must not exceed the moment capacity of the cross-section about the major-axis M_{cx}.

M_b is determined as the product of the bending strength p_b and the section modulus appropriate to the class of cross-section (see 5.3.1). For the great majority of rolled-sections this will be the plastic modulus S_x. Values of p_b are given in *Table 16* in terms of the equivalent slenderness λ_{LT}, which is defined as

$$\lambda_{LT} = \sqrt{\frac{\pi^2 E}{p_y}} \sqrt{\frac{M_p}{M_E}} \tag{5.2}$$

that is, the product of the quantity used as the abscissa in Figure 5.3 and a constant for a given grade of steel. The limiting values of λ_{LT} for which p_b may be taken as p_y, leading to $M_b = M_p$, have been extracted from *Table 17* and are presented in Table 5.2 as λ_{LO}. Providing lateral bracing is employed at a spacing not exceeding λ_{LO}, no allowance for failure by lateral-torsional buckling is necessary.

Inclusion of the factor m_{LT} in equation 5.1 recognizes the fact that uniform, single curvature moment is the most severe arrangement in terms of lateral stability. For this case m_{LT} is unity.

Alternatively, for the arrangement of Figure 5.4 in which the beam ABCD is loaded only at points of effective lateral restraint, producing an unrestrained length subjected only to unequal end moments, a reduced value may be used by following the rules of *Cl. 4.3.6.6* with m_{LT} being obtained from *Table 18*.

An appropriate formula that produces values at m_{LT} very close to those of *Table 18* is

$$m_{LT} = 0.57 + 0.33\beta + 0.10\beta^2 \geq 0.43 \tag{5.3}$$

in which β is the ratio of smaller to larger end moment on the segment between points of lateral restraint such that $1 \geq \beta \geq -1$.

Table 5.2   Values of maximum slenderness λ_{LO} for which beam strength is not influenced by lateral–torsional instability and $p_b = p_y$

p_y (N/mm²)	245	265	275	325	340	355	415	430	450
λ_{LO}	37	35	34	32	31	30	29	28	28

This special provision is based on the observation that results for moment gradient loading plot progressively higher on the frame of Figure 5.3 as M_1/M_2 decreases from 1.0 (single curvature) to -1.0 (double curvature). Thus, λ_{LT} is always calculated on the basis of uniform moment ($\beta = 1.0$) and the allowance for the actual shape of the moment diagram is made by conducting the design check of equation (5.1) using an 'equivalent uniform moment'.

Determination of the value λ_{LT} is most conveniently undertaken by using the formula of *Cl. 4.3.6.7*, viz.

$$\lambda_{LT} = uv\lambda \tag{5.4}$$

in which $\lambda = l/r_y$ is the minor axis slenderness
$u = 0.9$ for rolled sections see (*Cl. 4.3.6.8*)
$v =$ slenderness factor obtained from *Table 19*

For transverse loads applied between points of lateral restraint in such a way that is the beam tends to buckle so the load moves with the beam, a set of values for m_{LT} is provided to cover the more common arrangements. It is, of course, always safe to adopt the simplification of taking m_{LT} as unity.

In determining v, use is made of the 'torsional index' x; providing u is taken as 0.9, x may be approximated by the ratio of the overall depth to mean flange thickness D/T.

The procedure of equation (5.4) is effectively a way of bypassing the

Box girder bridge construction.

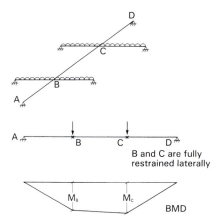

B and C are fully
restrained laterally

Figure 5.4   Beam loaded at points of effective lateral restraint.

explicit calculation of M_p and M_E as required by equation (5.2) in order to produce a much shorter calculation. Although this sacrifices something in accuracy, the effect on the final design is normally likely to be insignificant. Where accurate calculations are required, i.e. 'exact' values of u and x are needed for equation (5.4), *Appendix B* gives the full formulae; 'exact' values for standard rolled sections are listed in section tables [8].

Allowance for end supports which provide some measure of rotational restraint in the buckling plane is treated in *Cl. 4.3.5.1–4.3.5.3* which gives a set of effective length factors to be used when calculating λ. A second set of effective length factors is provided in *Cl. 4.3.5.4* and *4.3.5.6* for dealing with cantilevers [7, 9]. In both cases the 'destabilizing' load case corresponds to the situation in which a vertical load is applied to the top flange in such a way that it is free to move sideways as the beam tends to buckle in a lateral–torsional manner. As Figure 5.5 shows, such loads produce an

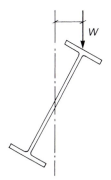

Figure 5.5   Torsion produced by top flange destabilizing load.

additional torsional effect leading to a reduction in the beam's lateral stability.

Example 5.1

Determine the buckling resistance moment for a $254 \times 146 \times 31$ UB in S275 steel assuming the beam to be laterally unsupported over a 3 m span.

Solution
From section tables, $r_y = 3.19$ cm, $D/T = 29.1$, $S_x = 394.8$ cm³

$$\lambda = l/r_y = 3000/31.9 = 94.0$$

Taking $x = D/T = 29.1$ gives $\lambda/x = 94.0/29.1 = 3.23$
From *Table 19*, noting that $N = 0.5$, $v = 0.900$
Taking $u = 0.9$ gives $\lambda_{LT} = 0.9 \times 0.9 \times 94.0 = 76$
From *Table 16*, for $p_y = 275$ N/mm² and $\lambda_{LT} = 76$, value of $p_b = 174$ N/mm²
$M_b = 174 \times 394.8 \times 10^3$ n/mm = <u>68.7 kN m</u>

Thus, for this example, lateral buckling reduces the bending resistance by $(275 - 174)/275 = 0.37$, one third, or turning the problem around, the maximum laterally unbraced span for which the full bending resistance $(M_p = S_x \times p_y)$ can be achieved is about 1.35 m (corresponding to a value of $\lambda_{LO} = 35$).

Example 5.2

Select a suitable UB section for the main beam of the structural arrangement shown in Figure 5.6 assuming the use of S275 steel.

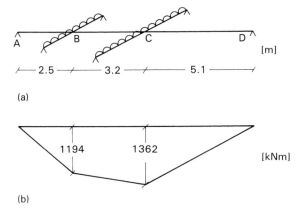

(a)

(b)

Figure 5.6 Beam of Example 5.2: (a) loading and support conditions; (b) bending-moment diagram.

Solution

The bending-moment diagram is shown in Figure 5.6(b). Noting that the two cross-beams provide full lateral restraint at B and C the design will be governed either by segment BC or by segment CD.

BC $\beta = 1194/1362 = 0.88$ and from *Table 18*, $m_{LT} = 0.94M$

$$\overline{M} = 0.94 \times 1362 = 1280 \, kN \, m$$

Using the procedure of Example 5.1 the lightest section capable of carrying 1280 kN m over a 3.2 m laterally unsupported span is a $762 \times 267 \times 173$ UB for which $M_b = 1475$ kN m.

CD $\beta = 0/1362 = 0.0$ and from *Table 18*, $m_{LT} = 0.57$

$$\overline{M} = 0.57 \times 1362 = 776 \, kN \, m.$$

Using the procedure of Example 5.1, for a moment of 776 kN m on a span of 5.1 m the $762 \times 267 \times 173$ UB is safe, since $M_b = 1072$ kN m.

Thus the design is controlled by the lateral stability of segment BC and the chosen section is a $762 \times 267 \times 173$ UB.

This example illustrates the use of the equivalent uniform moment concept when checking the strength of a beam that consists of several segments in the lateral plane. Often in such cases it is not possible to identify the critical segment simply by inspection. However, it is worth noting how, for this example, not using the equivalent uniform moment concept would require the provision of a section capable of carrying a moment of 1362 kN m over a span of 5.1 m, which would necessitate the use of a $914 \times 305 \times 201$ UB with a corresponding increase in steel weight of 16%.

5.3 Design of built-up sections (plate girders)

For many structures all of the beams may be provided from among the standard range of rolled sections. However, from time to time situations will arise in which none of the available sections has sufficient capacity. Such problems occur normally when it is necessary to provide a long span and/or to support a particularly heavy load, one frequently encountered example being the gantry girders provided in industrial buildings to carry the rails for a large-capacity overhead travelling crane. The normal solution is to use a built-up section, commonly called a plate girder, the proportions of which may be tailored specially to suit the design requirements. Nowadays it is normal practice to fabricate such sections simply by welding together three plates. However, in the past plate girders were often constructed by riveting or bolting, necessitating the use of angles to make the web-to-flange joints; several examples of this form of construction may still be seen. Different forms of plate girders are illustrated in Figure 5.7.

Because the designer has considerable freedom in proportioning a plate girder it is necessary for him to consider several structural problems which do not require the same attention when rolled sections are used. The most important of these are local buckling of the compression flange and shear buckling of the web [10]. Since the efficiency of the cross-section in resisting in-plane bending requires that the majority of the material be placed as far as possible from the neutral axis, it follows that minimum material consumption is frequently associated with the use of a very thin web. However, if premature failure due to web buckling in shear is not to occur, then web stiffening by means of vertical stiffeners, horizontal stiffeners or a combination of the two will normally be required [10]. In practice, the choice between a thin web provided with stiffeners or a thicker web requiring no stiffening (and therefore involving lower fabrication costs) depends upon a careful examination of the full costs of both forms of construction. Although flange capacity must also be checked, it is unusual for conventional plate girders to require compression stiffeners. On the other hand the ability of a slender web to resist both vertical buckling and/or local crushing often proves to be inadequate without the assistance of suitable stiffening.

Since the design of a plate girder requires consideration of the possibility of buckling of the flanges and/or the web, it is convenient to consider the various possible cases in terms of interaction as shown in Figure 5.8.

1   For zero shear, moment capacity is simply the product of the design strength of the material times the appropriate section modulus.
2   If the web is assumed to carry shear only, i.e all moment is assumed to be taken by the flanges, Figure 5.8a.
3   Providing the applied shear does not exceed 50% of the shear resis-

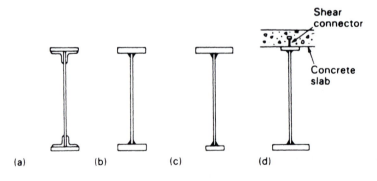

Figure 5.7   Plate girder types: (a) with flange angles; (b) welded; (c) unequal flanges; (d) composite.

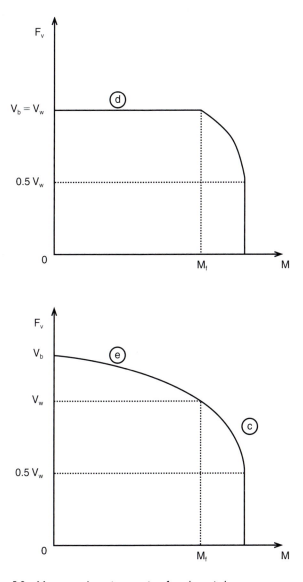

Figure 5.8 Moment–shear interaction for plate girders

tance of the web V_w, then the full moment capacity of the section, obtained as in 1 above, may be developed.

4   For combinations of shear greater than 0.5 V_w and moment greater than M_f, interaction should be considered as indicated by the region marked c in Figure 5.8b.

5.3.1   Local buckling effects in beams

The problem of local buckling in beams differs from that encountered in connection with columns (Section 4.1.2) chiefly because of the greater variety of stress conditions present in the component plates of a beam. Even in the case of a compression flange, the design condition could vary from a requirement that strains approaching yield be accommodated, to one in which strains greatly in excess of yield must be accepted with no reduction in strength. In addition, the web will be subject to some combination of shear and bending due to the overall flexural action and possibly also to additional local stresses in the immediative vicinity of point loads. Thus it becomes necessary to check for each of the following forms of instability:

1   buckling of the compression flange, noting carefully the level of strain which the design moment implies;
2   buckling of the web in shear and/or bending;
3   vertical buckling of a portion of the web under connected loads or over reactions.

(a)   Flange local buckling

Figure 5.9 gives examples of the two classes of plate element identified by *Cl. 3.5.1* of BS 5950: Part 1 as internal elements, which would correspond to the flange of a box beam, and outstand elements corresponding to the flange of the more commonly used I-section. For both types, four different ranges of 'compactness', each corresponding to a different performance requirement, are specified.

Class 1   Plastic $(b/t < \beta_1)$. Able to attain yield with sufficient plastic plateau to permit the redistribution of moments within the structure required for plastic design.
Class 2   Compact $(\beta_1 < b/t < \beta_2)$. Able to attain yield with sufficient plastic plateau to permit the section's full plastic moment to be attained.

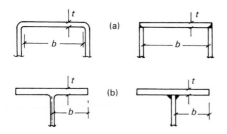

Figure 5.9   Types of plate element: (a) internal elements; (b) outstand elements. (*Dwight, Symposium on Revision of BS 449, 1978.*)

Class 3   Semi-compact ($\beta_2 < b/t < \beta_3$). Able to attain yield but local buckling limits available plastic plateau so that the section's full plastic moment cannot be attained.

Class 4   Slender ($b/t > \beta_3$). Local buckling prevents the attainment of the material design strength.

A further distinction is made between welded and non-welded elements on account of the more severe effects of the locked-in residual stresses present in the former [11]. For S275 and S355 steel the β limits of *Table 11* translate into the *b/t* limits given in Table 5.3. The moment capacity M_c of each of the four classes of section defined above is therefore calculated as:

1	plastic	$M_c = Sp_y$	
2	compact	$M_c = Sp_y$	(5.5)
3	semi-compact	$M_c = Zp_y$	
4	non-compact	$M_c < ZP_y$	

where S and Z are the plastic and elastic section moduli respectively. Thus for non-compact sections the moment capacity must be reduced according to the geometrical proportions of the section.

A special procedure is provided in *Cl. 3.5.6* for beams with compact flanges but semi-compact webs. This treats the cross-section as if it were plastic or compact in that rectangular stress blocks are employed when calculating the moment capacity but the compressive region of the web is taken as two areas of depth $20t\epsilon$ each as shown in Figure 5.10

The general procedure for beams of slender cross-section also involves the use of an effective section, in which the contribution to M_c of parts of the slender plate elements is neglected.

The idea is well supported both by rigorous theory and by observations of the behaviour of compressed plating in tests. These show the relationship between effective width and actual width to be dependent

Table 5.3   Limiting *b/t* values for plate elements subject to compression due to moment

	Internal element		Outstand element	
p_y values (N/mm²)	275	355	275	355
Non-welded β_1	28	25	9.0	7.9
β_2	32	28	10.0	8.8
β_3	40	35	15.0	13.2
Welded β_1	28	25	8.0	7.0
β_2	32	28	9.0	7.9
β_3	40	35	13.0	11.4

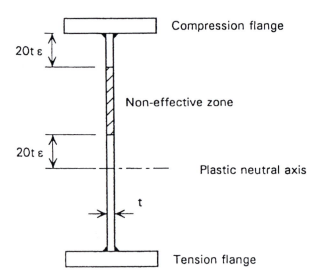

Figure 5.10   Effective cross-section for semi-compact web.

principally upon the plate thinness *b/t*, the conditions of support along the longitudinal edges (internal or outstand element) and the severity of residual stress (welded or non-welded). Thus the moment capacity of a beam containing a non-compact compression flange must be calculated using the proportions of the effective cross-section as shown in Figure 5.11. These correspond to the limits for semi-compact behaviour, i.e. any material in excess of the β_3 limit is ignored when calculating the section modulus *Z*.

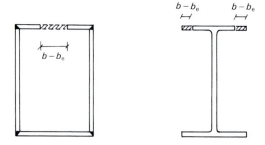

Figure 5.11   Effective sections for determining the section modulus of members containing slender plate elements. *(Dwight, Symposium on Revision of BS 449, 1978.)*

Example 5.3

Check whether the moment capacity of a welded plate girder comprising two 650 × 25 mm flange plates and one 1500 × 15 mm web plate will be affected by flange local buckling, assuming (a) S265 steel of design strength p_y = 265 N/mm², and (b) S345 steel of design strength p_y = 345 N/mm².

Solution

(a) For p_y = 265 N/mm², from *Table 11* maximum outstand b/t for flange to be compact = 9.0.

Actual b/t, using *Figure 5* = (325 − 15/2)/25 = 12.7, and $M_c < M_p$

Maximum b/t for flange to be semi-compact = 13

∴ section is semi-compact and $M_c = Zp_y$

I_x = (65 × 155³ − 63.5 × 150³)/12 = 2 311 614.6 cm⁴
Z_x = 2 311 614.6/(75 + 2.5) = 29 827.3 cm³
M_c = 265 × 29 827 × 10³ = 7904 kN m
S_x = 33 625 cm²

∴ reduction in capacity from that corresponding to compact

behaviour $= \left(\dfrac{9247 - 7904}{9247} \right) = 14.5\%$

(b) For p_y = 345 N/mm², maximum b/t for flange to be semi-compact = 11.7.

∴ section is slender and assume b_e is limit for semi-compact behaviour effective flange width b_e = 11.7 × 25 = 292 mm giving the effective section shown in Figure 5.12.

Locate neutral axis by taking moments about the top edge as 793 mm from top edge.

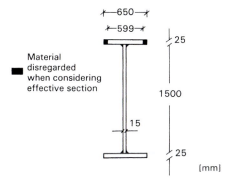

Figure 5.12   Plate girder of Example 5.3.

$$I_x = (599 \times 25)(793 - 12.5)^2 + (15 \times 1500^3/12) + (15 \times 1500)18^2$$
$$+ (650 \times 25)(757 - 12.5)^2$$
$$= 2.236 \times 10^{10} \text{ mm}^4$$
$$Z_x = (\text{top flange}) = 2.236 \times 10^{10}/793$$
$$= 2.82 \times 10^7 \text{ mm}^3$$
$$M_c = 345 \times 282 \times 10^7$$
$$= \underline{9729 \text{ kN m}}$$

∴ reduction in capacity from that corresponding to semi-compact

$$\text{behaviour} = \left(\frac{10\,290 - 9729}{10\,290} \right) = 5.5\%$$

Since the 'excess' material at the tips of the flanges cannot be included in calculations of the beam's moment capacity, consideration might be given to using a section which just meets the semi-compact requirements. This would avoid the complication of locating the neutral axis of the (effective) monosymmetric section.

(b)   Web behaviour
Girder webs will normally be subjected to some combination of shearing and bending stresses and, since the most severe condition in terms of web buckling is normally the pure shear case, it follows that it is those regions adjacent to supports of in the vicinity of point loads which generally control the design. Shear buckling occurs largely as a result of the compressive stresses acting diagonally within the web, as shown in Figure 5.13, with the number of waves tending to increase with

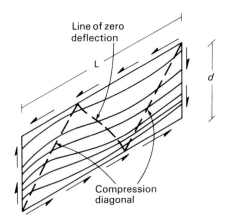

Figure 5.13   Buckling of a girder web in shear. (*After reference 6.*)

an increase in the panel aspect ratio a/d. The elastic critical stress q_c may be expressed as:

$$q_c = \left[0.75 + \frac{1}{(a/d)^2}\right]\left[\frac{1000}{d/t}\right]^2 \qquad \text{for } a/d \leqslant 1 \qquad (5.6)$$

$$q_c = \left[1 + \frac{0.75}{(a/d)^2}\right]\left[\frac{1000}{d/t}\right]^2 \qquad \text{for } a/d \leqslant 1$$

Because of the importance in equation (5.6) of the plate aspect ratio, shear buckling resistance may conveniently be improved by dividing the web into a series of panels by using intermediate vertical stiffeners. Examination of equation (5.6) suggests that a stiffener spacing which leads to panels having an aspect ratio a/d of between 0.5 and 2 will normally prove the most efficient. Although it is also possible to improve web strength by using horizontal stiffeners, this topic is not covered by BS 5950: Part 1 (which simply refers the reader to the bridge code BS 5400: Part 3) and is therefore beyond the scope of this text. Equation (5.6) forms the basis of the design method for webs provided in *Cl. 4.4.5.2* of BS 5950: Part 1, which gives the shear buckling resistance of a web as

$$V_w = dtq_w \qquad (5.7)$$

where q_w = shear buckling strength of the web

For stocky webs with $d/t \leqslant 62\epsilon$ q_w is simply the yield stress in shear, conveniently rounded to $0.6p_y$.

Values of q_w directly in terms of p_y, a/d and d/t are provided in *Table 21*.

Experiments (10, 12, 13] show that, providing sufficiently heavy stiffeners are employed, the web will be capable of withstanding loads in excess of the elastic buckling load. This occurs as a result of 'tension field action' in which the diagonal web tensile stresses act with the transverse stiffeners and the flanges to transfer the additional load by means of a truss type of action as shown in Figure 5.14. Ultimate load is not then reached until after the tension field has yielded at a load given approximately by

$$V_b = V_w + V_f \qquad (5.8)$$

in which V_w is given by equation (5.7) and V_f is the flange dependent shear buckling resistance.

BS 5950: Part 1 permits the use of this 'basic tension field action' for all girders other than crane gantry girders, provided certain conditions are met. The most important of these is that the end panels are made sufficiently strong to anchor the longitudinal force set up by the tension field. Rules for the detailed design of end panels are given in *Cl. 4.4.5.4*.

(a)

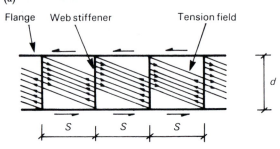

(b)

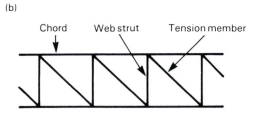

(c)

Figure 5.14   Tension field action in plate girder webs: (a) test girder showing well-developed tension fields (*H.R. Evans*); (b) load-carrying mechanism of tension fields, web stiffeners and flanges; (c) equivalence to behaviour of a truss. (*After reference 6.*)

The value of V_f in equation (5.8) should be obtained from:

$$V_f = \frac{P_v(d/a)[1 - (f_f/p_{yf})^2]}{1 + 0.15(M_{pw}/M_{pf})} \qquad (5.9)$$

in which f_f is the mean longitudinal stress in the smaller flange due to moment and/or axial load;

M_{pf} is the plastic moment capacity of the smaller flange, about its own equal area axis perpendicular to the plane of the web;

M_{pw} is the plastic moment capacity of the web, about its own equal area axis perpendicular to the plane of the web;

p_v is the shear capacity from 4.2.3;

p_{yf} is the design strength of the flange;

The difference between using equations 5.7 and 5.8 corresponds to the difference between the horizontal region marked d in Figure 5.8a and the convex region e of Figure 5.8b.

If region c is to be utilized, then reference should be made to *Cl. H.3*. This clause may also be required if, for a girder subject to some axial load, part of that axial load has to be resisted by the web. The simpler alternative is to arrange for any axial load to be taken by the flanges, in which case the above methods may still be used.

Example 5.4

The girder of Example 5.3 is required to carry a maximum shear of 3000 kN. Assuming that tension field action is not to be utilized in the design, determine whether intermediate stiffening is necessary. Take the design strength of the steel p_y as 275 N/mm². How thick must the web be made in order that this same load can be carried without the need for intermediate stiffeners?

Solution

From equation (5.7), $V_w = dtq_w$

Using d/t of $1500/15 = 100$ in *Table 21* gives, for no stiffeners ($a/d = \infty$), $q_w = 116$ N/mm²

$$\therefore V_w = 1500 \times 15 \times 116 = 2\,610\,000 \text{ N}$$
$$= \underline{2610 \text{ kN}}$$

Therefore stiffening is required

Required $q_w = 300 \times 10^3/1500 \times 15 = 133$ N/mm²

From *Table 21*, for $d/t = 100$, max. a/d corresponding to this strength $= 1.5$

$$\therefore \text{ provide stiffeners at } 1.5 \times 1500 = \underline{2250 \text{ mm intervals}}$$

For second part of this example a trial-and-error approach is necessary since V_w depends on q_w which is itself dependent on t. Clearly t must be greater than 15 mm.

Try $t = 18$ mm $\rightarrow d/t = 83.3$ and q_w (for $a/d = \infty$) $= 138$ N/mm²

$$\therefore V_w = 1500 \times 18 \times 138 = \underline{3726 \text{ kN}}$$

Because the web in this example is not particularly slender ($d/t \leqslant 100$) the better solution is probably to increase its thickness and avoid the need for stiffening. However, inspection of *Table 21* shows that for deeper girders comparatively much larger strength increases result from the use of stiffeners, particularly closely spaced stiffeners. For example, for $d/t = 250$, stiffeners at $0.5d$ double the shear strength while stiffeners at $0.4d$ produce at least a six-fold improvement.

Example 5.5

Assuming a stiffener spacing equal to the panel depth, determine the shear capacity of the girder of Example 5.3.

Solution
Cl. 4.4.5.3 gives:

$$V_b = V_w + V_f$$

and *Cl. 4.4.5.2* gives:

$$V_w = dtq_w$$

from *Table 21* for $d/t = 100$ and $a/d = 1.0$, $q_w = 147$ N/mm²

$$\therefore V_w = 1500 \times 15 \times 147 = 3308 \text{ kN}$$

from *Cl. 4.4.5.3*

$$V_f = \frac{P_v(d/a)[1 - (f_f/p_{yf})^2]}{1 + 0.15(M_{pw}/M_{pf})}$$

$$M_{pf} = 275 \times 650 \times 25^2/4 \quad = 2730 \text{ kN m}$$
$$M_{pw} = (750 \times 15)275 \times 750 = 2320 \text{ kN m}$$

From *Cl. 4.2.3*

$$P_v = 0.6p_y A_v$$
$$= 0.6 \times 275 \times 1500 \times 15$$
$$= 3713 \text{ kN}$$

Clearly contribution of V_f to V_b depends on level of direct stress f_f due to bending in the flange.

For minimum contribution $f_f = p_{yf}$, and

$$V_f = \frac{3713(1.0)[1 - 1]}{1 + 0.15(2320/2730)} = 0 \text{ kN}$$

and $V_b = V_w = \underline{3308 \text{ kN}}$

For maximum contribution $f_f = 0$, and

$$V_f = \frac{3713(1.0)[1 - 0]}{1 \times 0.15(2320/2730)} = 3293 \text{ kN}$$

∴ maximum possible shear resistance (in the absence of any coincident bending) is given by

$$V_b = 3308 + 3293$$

$$= \underline{6601 \text{ kN}}$$

The shear resistance of the panel at a location between the points of minimum and maximum coincident moment considered above will clearly fall part-way between these two limits on V_b, depending on the local value of f_f.

Following Example 5.5 it is also clearly possible to select numerous alternatives in which part of the shear resistance is provided by the flanges; this will clearly require larger flanges (since they cannot be fully stressed in bending) but will lead to a lighter web.

Design of transverse stiffeners
Transverse stiffeners must be proportioned so as to satisfy two conditions:

1   They must be sufficiently stiff not to deform appreciably as the web tends to buckle.
2   They must be sufficiently strong to withstand the shear transmitted by the web.

Since it is quite common to use the same stiffeners for more than one task (for example the stiffeners provided to increase shear buckling capacity can also be used as load-bearing stiffeners to assist the web in carrying heavy point loads,), the above conditions must also, in such cases, include the effects of any additional direct loading.

Condition (1) is covered by *Cl. 4.4.6.4* by requiring web stiffeners to have a second moment of area at least equal to

$$I_s \geq 0.75dt^3_{\text{min}} \qquad \text{for } a/d \geq \sqrt{2} \qquad\qquad (5.10)$$
$$I_s \geq 1.5(d/a)^2 dt^3_{\text{min}} \text{ for } a/d < \sqrt{2}$$

in which t_{min} = minimum required web thickness for the actual stiffener spacing, these values being increased in accordance with *Cl. 4.4.6.5* when lateral forces and/or eccentrically applied transverse loads must also be carried by the stiffener. The strength requirement is checked by ensuring that the stiffener acting as a strut is capable of withstanding F_q, the difference between the shear actually present adjacent to the stiffener and the shear capacity of the (unstiffened) web, together with any coexisting reaction or moment. Since the portion of the web immediately adjacent to the stiffener tends to act with it, this 'strut' is assumed to consist also of a length of web of 20t on either side of the stiffener centre-line giving an effective section in the shape of a cruciform. Full details of this strength check are given in *Cl. 4.4.6.6*. If tension field action is being utilized then the stiffeners bounding the end panel must also be capable of accepting the additional forces associated with anchoring the tension field.

Example 5.7

Design a suitable vertical stiffener for the stiffened version of the girder of Example 5.4.

Solution
Since $a/d = 1.0$, use second expression in (5.10) to give

$$I_s \geq 1.5 \ (d/a)^2 dt^3_{min}$$
$$\geq 1.5(1)^2 \times 1500 \times 15^3 = 759 \ cm^4$$

Assuming the use of double-sided stiffeners of (say) 15 mm plate, since

$$I_s = \frac{15(2b)^3}{12}$$

gives $b = [12 \times 7\ 590\ 000/120]^{\frac{1}{3}} = 91$ mm

$\therefore$ use a pair of 100×15 mm plates
   check strength using *Cl. 4.4.6.6*
   $V = 3000$ kN $V_w = 2610$ kN
$\therefore F_q = 3000 - 2610 = 390$ kN

Effective width of plate $= 20 \times 15 = 300$ mm

$$I_x = 125 \ cm^4 \qquad A = 75 \ cm^2 \qquad r_x = 12.9 \ mm$$

Take effective length l as $d = 1500$ mm (assumes no lateral restraint to flanges at stiffener position, *Cl. 4.5.2.3*).

$\therefore \lambda = 1500/12.9 = 116$

From *Table 24* for $p_y = 275$ N/mm^2 $\qquad p_c = 102$ N/mm^2

$\therefore P_q = 102 \times 7500$ N $= \underline{765 \ kN}$

Since this exceeds F_q, stiffener has adequate strength.

(c)   Web buckling due to vertical loads

The application of heavy concentrated loads to a girder will produce a region of very high stress in the part of the web directly under the load. One possible effect of this is to cause outwards buckling of this region rather as if it were a vertical strut with its ends restrained by the beam's flanges. This situation also exists at the supports where the 'load' is now the reaction and the problem is effectively turned upside down. It is usual to interpose a plate between the point load and the beam flange, whereas in the case of reactions acting through a flange this normally implies the presence of a seating cleat. In both cases, therefore, the load is actually spread out over a finite area by the time it passes into the web as shown in Figure 5.15. This is referred to as 'dispersion into the web' and is con-trolled largely by the dimensions of the plate used to transfer the load, which is itself termed 'the stiff length of bearing'.

Because it is virtually impossible to provide anything approaching a rig-orous theoretical treatment of this problem, design methods are based normally upon empirical formulae derived directly from tests. Thus *Cl. 4.5.2.1* of BS 5950: Part 1 assumes the load to be carried by a vertical strut, the width of which is dependent upon the stiff length of bearing pro-vided. This leads to the following expression for web buckling strength:

$$P_w = (b_1 + n_1)tp_c \qquad (5.11)$$

in which b_1 = stiff length of bearing given by *Cl. 4.5.1.3*
$\quad n_1$ = length obtained by dispersion of 45° through half the depth of the section
$\quad t$ = web thickness
$\quad p_c$ = compressive strength according to curve *c*

It is usual to assume that both flanges provide full rotational restraint to this 'strut' in which case λ may be taken as $2.5d/t$ corresponding to a strut

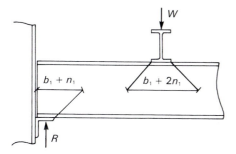

Figure 5.15   Dispersion of concentrated loads and reactions.

effective length of $0.7d$. However, in situations where movement of one flange relative to the other is possible, larger slendernesses are appropriate, as explained in *Cl. 4.5.3.1*.

It will often be the case that an otherwise satisfactory girder will prove to have inadequate strength according to equation (5.11). One remedy is to employ load-bearing stiffeners to carry the excess load. Indeed this problem is encountered so frequently that designers will often call for such stiffeners at load and reaction points as a matter of course. Moreover it is not confined to built-up girders; many UB sections have webs that will be found to be inadequate when checked against equation (5.11).

The design of load-bearing stiffeners is essentially the same as the design of vertical stiffeners for strength, as explained in the previous section. The load is again assumed to be resisted by a strut comprising the actual stiffeners plus a length of web of $20t$ on either side, giving an effective cruciform section. Providing the loaded flange is laterally restrained the effective length of this 'strut' may be taken as $0.7L$. Although no separate stiffness check is necessary, load-bearing stiffeners must be of sufficient size that if the full load were to be applied to them acting independently, i.e. on a cross-section consisting of just the stiffeners, then the stress induced should not exceed the design strength by more than 25%.

The exact functions of the different types of web stiffener that might be required on a slender web are explained in *Cl. 4.5.1*, which also refers the reader to the sections of BS 5950: Part 1 that should be considered for the design of each type.

Example 5.8

For the girder of Example 5.3 check whether a 3000 kN reaction can be carried, assuming it to act through a cleat of 15 mm thickness.

Solution

From equation (5.11) $P_w = (b_1 + n_1)tp_c$

From *Cl. 4.5.1.3* $b_1 \ = 2 \times (15 + 25) = 80$ mm

From *Cl. 4.5.2.1* $n_1 \ = d/2 = 750$ mm

From *Cl. 4.5.3* $\lambda \ = 2.5 \times 1500/15 = 250$

Using *Table 24* $p_c \ = 28$ N/mm^2

$\therefore P_w = (80 + 750)15 \times 28 \times 10^{-3} = 349$ kN

and web stiffeners are required.

Assuming $p_c = 200$ N/mm^2 (λ is likely to be low), required area of strut comprising stiffener + attached plating $= 300 \times 10^3/200 = 15\,000$ mm^2

If using double-sided stiffeners of 20 mm plate, stiffener width needs to be

$\frac{1}{2}(1500 - (2 \times 300 \times 15))/20 = 150$ mm

∴ try 150 × 20 mm stiffeners

$$I_x = 20 \times 315^3/12 \qquad\qquad = 52.1 \times 10^6 \text{ mm}^4$$
$$A = 2(20 \times 150) + 600 \times 15 = 15\,000 \text{ mm}^2$$
$$r_x = 58.9 \text{ mm}$$

From *Cl. 4.5.3.1* $L_E = 0.7d$

∴ $\lambda = 0.7 \times 1500/58.9 = 17.8$ and from *Table 24* $p_c = 263$ N/mm²
∴ $P_q = 15\,000 \times 263$ N = <u>3945 kN</u>

Bearing strength of webs

A second possible form of failure for a web subject to a locally applied, high compressive load is through the development of unacceptably high bearing stresses at the junction with the loaded flange. This may be checked using *Cl. 4.5.2.1* and, if found necessary, local strengthening provided in the form of bearing stiffeners designed according to *Cl. 4.5.2.2*.

Exercises

1   Select a UB section capable of safely carrying a total uniformly distributed load of 170 kN over a span of 7.2 m, assuming the use of S275 steel and the provision of full lateral support to the beam.

[305 × 165 UB 40]

2   Determine the buckling resistance moment for a 457 × 152 UB 60 in S275 steel when it is simply supported over a span of 3.5 m.

[190 kN m]

3   Determine the buckling resistance moment for a 356 × 127 UB 33 in S275 steel for a span of 4.2 m, assuming that the applied loading produces moments which vary linearly from a maximum at one end to one quarter of this value at the other, both values being in a clockwise sense.

[92.5 kN m)

4   Select a UB in S275 steel capable of safely carrying end moments of 640 kN m and 128 kN m over a laterally unsupported span of 6.5 m assuming that the moments produce single curvature bending.

[610 × 229 UB 125]

5   What is the moment capacity of a short length of welded plate girder fabricated from two 600 × 30 mm flange plates and one 1600 × 12 mm web plate assuming S275 steel? What changes, if any, are required in plate thickness if the section is to be capable of carrying its full plastic moment?

[8943 kN m, $T = 35$ mm, $t = 18$ mm]

6 Determine the buckling resistance moment for a welded plate girder comprising 500 × 25 mm flange plates and a 1200 × 12 mm web plate in S275 steel assuming a laterally unbraced span of 6 m.

[4022 kN m]

7 A plate girder web is to be fabricated from a plate 1300 mm deep by 12 mm thick. Assuming S275 steel, determine at what spacing vertical stiffeners must be placed if the girder is to be capable of carrying a shear load of 1350 kN without the use of any flange contribution.

[2.6 m spacing]

8 Using the method of *Cl. 4.4.4*, design a plate girder in steel of approximately 1250 mm overall depth to withstand coincident moment and shear loads of 700 kN m and 2000 kN. Indicate the spacing of vertical stiffening, if any, necessary.

[Many solutions are possible but 700 × 35 mm flanges and a 1200 × 12 mm web with stiffeners at 1200 mm spacing would be satisfactory]

References

1 Nethercot, D.A., Salter, P.R. and Malik, A.S. (1989) *Design of Members Subject to Combined Bending and Torsion*, The Steel Construction Institute.

2 Neal, B.G. (1970) *Plastic Methods of Structural Analysis*, Chapman and Hall, London.

3 Horne, M.R. (1979) *Plastic Theory of Structures*, 2nd edn, Pergamon, Oxford.

4 Kirby, P.A. and Nethercot, D.A. (1979) *Design for Structural Stability*, Granada, St Albans, 1979.

5 Timoshenko, S.P. and Gere, J.M. (1961) *Theory of Elastic Stability*, 2nd edn, McGraw-Hill, New York.

6 Trahair, N.S. and Bradford, M.A. (1998) *The Behaviour and Design of Steel Structures to AS 4100,* E & FN Spon, London.

7 Nethercot, D.A. (1983) Elastic lateral torsional buckling, in R. Narayanan (ed.) *Stability and Strength of Beams and Beam-Columns*, Applied Science Publishers, London, pp. 1–34.

8 Steel Construction Institute (1997) *Steelwork Design Guide to BS 5950: Part 1: 1990. Volume 1, Section Properties, Member Capacities*, 4th edn.

9 Nethercot, D.A. (1973) The effective lengths of cantilevers as governed by lateral buckling, *The Structural Engineer*, **51**(5), 161–8.

10 Evans, H.R. (1984) Longitudinally and transversely reinforced plate girders, in R. Narayanan (ed.) *Stability and Strengths of Plated Structures*, Applied Science Publishers, London, pp. 1–38.

11 Dwight, J.B. and Moxham, K.E. (1969) Welded steel plates in compression, *The Structural Engineer*, **47**(2), 49–66.

12 Rockey, K.C., Evans, H.R. and Porter, D.M. (1981) The design of stiffened web plates – a state of art report, in K.C. Rockey and H.R. Evans (eds) *The Design of Steel Bridges*, Granada, London, pp. 215–42.

13 Rockey, K.C. and Skaloud, M. (1972) The ultimate load behaviour of plate girders loaded in shear, *The Structural Engineer*, **50**(1), 29–47.

Chapter 6

Members under combined axial load and moment

Chapters 3–5 have dealt with the design of members subjected to a single form of loading, such as tension, and bending about one axis. However, situations will often arise in which the loading on a member cannot reasonably be represented as a single dominant effect. Such problems require an understanding of the way in which the various structural actions interact with one another. In the simplest cases this may amount to nothing more than a direct summation of load effects. Alternatively for more complex problems, careful consideration of the complicated interplay between both the individual load components and the resulting deformations is necessary.

The design approach discussed in this chapter is intended for use in situations where a single member is to be designed for a known set of end moments and forces. As such it is applicable to members in 'simple construction' although, as will be explained in Chapter 10, similar approaches are also possible for certain framing arrangements which fall within the general classification of 'continuous construction'.

Because of the additional complexity due to buckling associated with compressive loads, it is convenient to deal with the cases of tension plus bending and compression plus bending separately.

6.1 Combined tension and moments

The procedures outlined previously in Chapter 3 for angle ties are valid only for those cases in which bending is produced solely by the fairly small eccentricities between the loaded leg and the member axis. For more general problems each load component must be considered separately since it is not known in advance which will be dominant.

The assumption of elastic behaviour leads to a simple design approach based on limiting the sum of the individual stresses at a cross-section to the design strength of the material p_y.

$$p_a + p_{bx} + p_{by} \not> p_y \tag{6.1}$$

in which p_a = axial stress due to load F

p_{bx} = maximum bending stress due to moments M_x about the $x - x$ axis

p_{by} = maximum bending stress due to moments M_y about the $y - y$ axis

Converting this to an expression for loads and rearranging, gives

$$\frac{F}{Ap_y} + \frac{M_x}{Z_x p_y} + \frac{M_y}{Z_y p_y} \ngtr 1 \tag{6.2}$$

in which Z_x = elastic section modulus about the $x - x$ axis

Z_y = elastic section modulus about the $y - y$ axis

It has already been explained in Chapter 5 how stocky beams of compact cross-section may be expected to develop their full plastic moment capacity $M_p = Sp_y$. Therefore, in order that (6.2) reduces in the limiting cases of $F \rightarrow 0$ to the design condition for beams, the quantities $Z_x p_y$ and $Z_y p_y$ should be replaced by M_{cx} and M_{cy}, the cross-sectional moment capacities obtained from equation (5.5) as explained in Chapter 5, to give

$$\frac{F}{A_e p_y} + \frac{M_x}{M_{cx}} + \frac{M_y}{M_{cy}} \ngtr 1 \tag{6.3}$$

in which A_e is the effective area (see Chapter 3).

Use of the major-axis bending cross-sectional strength in (6.3) means that no allowance is made for lateral–torsional buckling effects, i.e. by using M_b from equation (5.1) for M_{cx}. Although the presence of an axial tension may be expected to reduce any tendency towards instability, it would seem prudent in cases where F is small and M_b is significantly less than M_{cx} not to disregard this since, in the limiting case of $M_y = 0$ and $F \rightarrow 0$, (6.3) should agree with equation (5.1). In the absence of clear evidence to the contrary it is suggested that the same allowance be made for all values of axial tension and (6.3) be checked using M_b for M_{cx}; this may well be rather conservative in many instances.

Clause 4.8.2 of BS 5950: Part 1 uses (6.3) to check members at the points of maximum tension and bending; it suggests that this will usually be the ends. This is a linear interaction in which each of the three terms has equal effect. Figure 6.1 shows how it correctly tends towards the previously derived design conditions for the component cases as one form of loading becomes dominant.

More sophisticated analysis of this problem using the principles of plastic theory [1, 2] has shown that for compact cross-sections, i.e. those satisfying the geometrical limits of *Table 11* or *12* for no reductions in strength due to local buckling effects, (6.3) may be replaced by

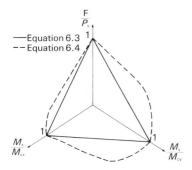

Figure 6.1   Interaction for strength under combined loading.

$$\left(\frac{M_x}{M_{rx}}\right)^{z_1} + \left(\frac{M_y}{M_{ry}}\right)^{z_2} \not> 1 \qquad\qquad (6.4)$$

in which M_{rx} = reduced moment capacity about the $x - x$ axis in the presence of the axial load F

M_{ry} = reduced moment capacity about the $y - y$ axis in the presence of the axial load F

z_1 = 2.0 for I- and H-sections and for solid or hollow circular
sections, 5/3 for solid and closed hollow rectangular sections and 1.0 in all other cases

z_2 = 1.0 for all sections other than solid and closed hollow rectangular sections for which a value of 5/3 may be used and
solid and closed circular hollow sections for which a value
of 2.0 may be used

Use of (6.4) will normally lead to higher results, as shown in Figure 6.1.
In using (6.4) the values of M_{rx} and M_{ry} for standard sections may be
obtained from section tables [3]. Alternatively the following expressions
from Appendix J may be used for rolled I- and H-sections.

$$S_{rx} = S_x - \left[\frac{A^2}{4t}\right]n^2 \text{ for } n \le t(D - 2T)/A \qquad\qquad (6.5)$$

$$S_{rx} = \left[\frac{A^2}{4B}\right]\left[\left[\frac{2BD}{A} - 1\right] + n\right](1 - n) \text{ for } n > t(D - 2T)/A$$

$$S_{ry} = S_y - \left[\frac{A^2}{4D}\right]n^2 \text{ for } n \le tD/A$$

$$S_{ry} = \left[\frac{A^2}{8T}\right]\left[\left[\frac{4BT}{A} - 1\right] + n\right](1 - n) \text{ for } n > tD/A$$

in which S_{rx}, S_{ry} = reduced plastic modulus in the presence of axial load F
$\qquad S_x$, S_y = plastic modulus for zero axial load
$\qquad n = F/Ap_y$

Values of plastic section moduli for angles bent about their rectangular axes are available [4]; for other types of cross-section, for example channels and fabricated I-sections, it is necessary to refer to texts on plasticity theory [1, 5]. In order that (6.4) be consistent with the procedures of Chapter 5 for simple bending, the values of M_{rx} and M_{ry} used should not exceed $1.2p_yZ_x$ and $1.2p_yZ_y$ respectively.

Example 6.1

Check whether a stocky $254 \times 146 \times 31$ UB of S275 steel is safe under (factored) loads $F = 340$ kN and $M_x = 85.0$ kN m.

Solution
Since the member is compact take $M_{cx} = p_bS_x$

From (6.3), $\dfrac{340 \times 10^3}{3990 \times 275} + \dfrac{85 \times 10^6}{394.8 \times 10^3 \times 275} = 0.310 + 0.783$

$\qquad\qquad\qquad\qquad\qquad = \underline{1.093}$, and section is not safe.

Using (6.4, 6.5), $n = \dfrac{340 \times 10^3}{3990 \times 275} = 0.310$

Change formula for S_{rx} in (6.5) at

$\qquad n = 6.0\,(251.4 - 2 \times 8.6)/3970 = 0.354$

$\qquad \therefore S_{rx} = 393\,000 - \left[\dfrac{3970^2}{4 \times 6.0}\right]0.31^2$

$\qquad\qquad = 329.9 \text{ cm}^3$

$\qquad \dfrac{M_x}{M_{rx}} = \dfrac{85 \times 10^3}{330 \times 275} = \underline{0.937}$ and section is safe

Alternatively using the formula from section tables [3]

$\qquad S_{rx} = 393 - 653.9 \times 0.31^2$
$\qquad\quad = 330 \text{ cm}^2$

Gives an identical result

This example shows how the use of progressively more 'exact' procedures leads to corresponding increases in the predicted capacity.

Example 6.2

If M_x is reduced to 60.0 kN m what moment may safely be applied about the minor axis.

Solution
From section tables, $S_y = 88.78$ cm^3

From (6.3), $\dfrac{340 \times 10^3}{3990 \times 275} + \dfrac{60 \times 10^6}{394.8 \times 10^3 \times 275} + \dfrac{M_y}{88.78 \times 10^3 \times 275}$

$$= 1.0$$
$$\text{gives } M_y = 3.36 \text{ kN m}$$

Using (6.4, 6.5) noting that $z_1 = 2$ and $z_2 = 1$
   Change formula for S_{ry} at $n = 6.0 \times 251.4/3970 = 0.38$

$$\therefore S_{ry} = 94\,100 - \left[\frac{3970^2}{4 \times 251.4}\right] 0.31^2$$

$$= 92.6 \text{ cm}^3$$

and $M_{ry} = 275 \times 92.6 = \underline{25.46 \text{ kN m}}$

$$\therefore \left(\frac{60}{90.75}\right)^2 + \left(\frac{M_y}{25.46}\right) = 1$$

and $M_y = \underline{14.33 \text{ kN m}}$
   If the formulae in section tables [3] are used to obtain M_{ry} an identical result is obtained.

Once again (6.4) gives a significantly higher result than (6.3), with the use of the larger M_r values obtained from section tables producing an identical result.

6.2   Combined compression and moments

When the axial component of the loading is compressive then the member's strength may be limited by either of the two conditions:

1   local capacity at the most heavily loaded cross-section;
2   overall buckling.

   The first of these is essentially equivalent to the problem discussed above, while the overall buckling of a beam column closely resembles column stability as discussed in Chapter 4. However, because the loading

Reinforcement for an opening in a beam web.

may take several different forms, so the member's response must be treated under a number of different headings.

The most common form of beam-column problem in building structures is the vertical member supporting (usually horizontal) beams; a typical example is shown in Figure 6.2. Because of the assumptions regarding connection behaviour associated with 'simple construction', the loading on the stanchion may be taken as that shown in Figure 6.3, i.e. an axial load F due to accumulated load from the floors above plus moments due to the beam reactions F_x and F_y assumed to act at known eccentricities e_y and e_x to the column faces. Guidance on the choice of suitable values for these eccentricities is provided in *Cl. 4.7.7* of BS 5950: Part 1. Thus, in the most

Figure 6.2 Typical arrangement of beams and columns in a multistorey building.

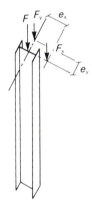

Figure 6.3 Loading on a beam column in 'simply designed' frame.

general case, the beam column is subject to compression plus moments about both axes. If the loading and/or the beam arrangement is different at different levels then these moments will not be the same at both ends, that is moment gradients will exist as shown in Figure 6.4. Of course, if some beams are absent or in the case where similar beams on opposite sides of the column carry identical loads so that the beam moments exactly

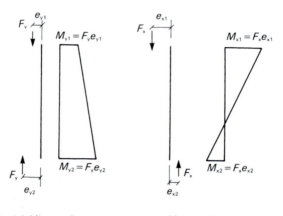

(a) Minor-axis
$\beta_y = M_{y1}/M_{y2}$ is positive

(b) Major-axis
$\beta_x = M_{x2}/M_{x1}$ is negative

Figure 6.4   Bending moments in a beam column. (a) Minor axis ($\beta_y = M_{y1}/M_{y2}$ is positive); (b) major axis ($\beta_x M_{x2}/M_{x1}$ is negative).

balance, then the loading may reduce to a simpler form. Three distinct cases may be identified as shown in Figure 6.5.

1   The thrust is applied with an eccentricity about the minor axis (or if the eccentricity is about the major axis then either the column is prevented from deflecting out of this plane, by properly designed cladding for example, or there is no tendency for out-of-plane buckling due to the applied moment, as happens when the member is a circular tube) in which case the member will collapse by excessive deformation in this plane.

2   The thrust is applied with an eccentricity about the major axis and the member fails by a combination of bending about the weak axis and twisting, similar to lateral–torsional beam buckling.

3   The thrust is applied with an eccentricity about both axes, in which case the member will collapse by combined bending and twisting.

Thus case (1) represents an interaction between column buckling and simple uniaxial beam bending, case (2) represents an interaction between column buckling and beam buckling, and case (3) represents the interaction of column buckling and biaxial beam bending. Clearly case (3) is the most general case with the others being more limited versions.

Not surprisingly the analytical background to the beam-column problem is extremely complex. The next three sections therefore provide only a relatively simple description; readers who are interested in obtaining a more complete understanding are advised to consult references [2, 6, 7].

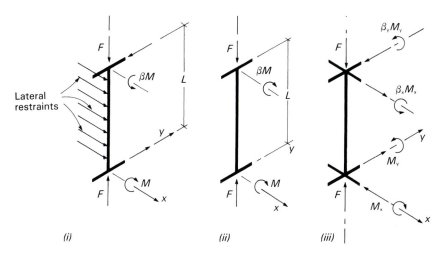

Figure 6.5   Three classes of beam column problem: (1) In-plane behaviour: column deflects v in $y - z$ plane only [$F + M_x$ with bracing; $F + M_y$]; (2) flexural–torsional buckling: column deflects v in $y - z$ plane, then buckles by deflecting u in $x - z$ plane and twisting $\varnothing$ [$F + M_x$]; (3) biaxial bending: column deflects u, v and twists $\varnothing$ [$F + M_x + M_y$].

6.2.1   Case (1): In-plane strength

Within the elastic range, case (1) of Figure 6.5 may be analysed using the basic Euler theory of compression members [2]. Assuming equal end moments M, as shown in Figure 6.6, and setting up and solving the resulting differential equation permits the deflected form and hence the bending

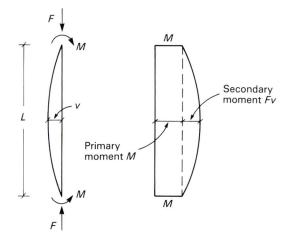

Figure 6.6   In-plane behaviour of beam column.

moments and stresses in the beam column to be determined. Because of the additional bending deformations caused by the compression F acting through an ever-increasing effective eccentricity (the lateral deformations v), the member will respond in a non-linear fashion to the applied loads as shown in Figure 6.7. The theoretical upper limit of F will be the elastic critical value $P_{cr} = \pi^2 EI/L^2$. However, this assumes indefinite elastic behaviour. If the stress due to compression

$$f_a = \frac{F}{A} \tag{6.6}$$

together with the maximum bending stress

$$f_b = \frac{M_{max}}{Z} \tag{6.7}$$

is limited to the material yield stress σ_y, noting also that as $M \to 0$ F must be limited to P_{cr}, then the corresponding values of F and M will be related by

$$\frac{F}{P_{cr}} + \frac{M}{M_y}\left(\frac{M_{max}}{M}\right) = 1.0 \tag{6.8}$$

in which (M_{max}/M) allows for the additional secondary moments due to deformation. Rather than use the exact expression for (M_{max}/M), it is convenient to replace this with the close approximation

$$(M_{max}/M) \simeq \frac{M}{M_y(1 - F/P_{cr})} \tag{6.9}$$

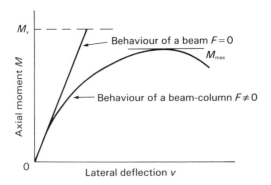

Figure 6.7   Non-linear response of a beam column assuming elastic behaviour.

This quantity is often termed an 'amplification factor' since it amplifies the primary moment M to give the total moment (primary + secondary). Thus equation (6.8) becomes

$$\frac{F}{P_{cr}} + \frac{M}{M_y(1 - F/P_{cr})} = 1.0 \tag{6.10}$$

At low slendernesses, when P_{cr} will be so large that the amplification factor will have negligible effect, it plots as a straight line interaction between the axial (F/P_{cr}) and bending (M/M_y) effects. However, as slenderness increases so the effects of secondary bending becomes more significant, resulting in an increasingly concave interaction as shown in Figure 6.8.

More rigorous analysis of this problem [2] allowing for the effects of yielding, residual stress, initial lack of straightness, etc., namely all those factors present in the behaviour of real steel members as discussed in Chapter 4, shows that the actual strength of beam columns may be quite closely predicted using a modified version of equation (6.10). Thus *Cl. 4.8.3.3.2* of BS 5950: Part 1 uses

$$\frac{F}{P_c} + \frac{1}{2}\frac{F}{P_c}\frac{M}{M_c} + \frac{M}{M_c} = 1 \tag{6.11}$$

in which P_c and M_c are the uniaxial compressive and bending strengths and the product term approximates the function of the amplification factor.

(a) Effect of non-uniform moment

Returning to elastic analysis and Figure 6.6, if the end moments are now taken as M and βM, where $1 \geqslant \beta \geqslant -1$ and $\beta = 1$ corresponds to uniform single curvature bending, it may be shown [2] that the first yield

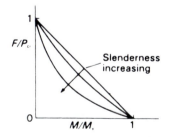

Figure 6.8 Elastic limit strength for in-plane behaviour of beam columns in uniform bending according to equation (6.10).

interaction adopts the form of Figure 6.9. As the applied moments tend towards double curvature ($\beta \to -1$), so the primary and secondary bending effects become less directly additive, with the result that the interaction plots higher. Eventually a situation will be reached in which yield occurs first at one end under the action of the primary moment alone, corresponding to the intersection with the zero slenderness (strength) interaction boundary. It is possible to represent these results quite accurately using equation (6.10), providing an equivalent value mM is used. Coincidentally the relationship between m and β is very similar to that introduced in Chapter 5 for dealing with the lateral–torsional buckling of beams and a suitable expression is:

$$m = 0.57 + 0.33\beta + 0.10\beta^3 \not< 0.43 \tag{6.12}$$

Since equation (6.11) now checks overall stability of the member, it is, of course, also necessary to ensure against local overstressing at the more heavily loaded end using the full value of the moment, i.e. to keep within the upper boundary of Figure 6.9.

Based on the results of rigorous theoretical studies [2] together with test data it has been found that the 'equivalent uniform moment' concept may be used with the design expression of equation (6.11). Thus M may now be reinterpreted as mM. In such cases it is necessary to check local strength separately using (6.3) or (6.4).

Example 6.3

What is the axial load capacity of a 203×203 UC 60 of 3.1 m height assuming that the loading acts at an effective eccentricity of 100 mm in the $y - y$ direction at both ends (assume $p_y = 275 \text{ N/mm}^2$).

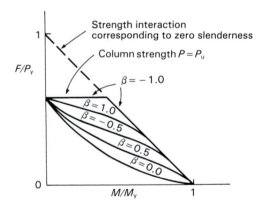

Figure 6.9   Effect of moment gradient β on elastic limit interaction.

Solution

From section tables, $A = 75.8$ cm^2, $r_y = 5.19$ cm, $Z_y = 199.0$ cm^3

$\therefore$ $L/r_y = 3100/51.9$ and, noting from *Table 23* that strut curve *c* is appropriate corresponding value of p_c from *Table 24* = 200 N/mm^2

$\therefore$ $\quad P_c = 200 \times 75.8 \times 10^{-1} = 1516$ kN

$\quad M_{cy} = 1.2 \times 275 \times 199.0 \times 10^{-3} = 65.7$ kN m

and from equation (6.11) $\dfrac{F}{1516} + \dfrac{F \times 0.1}{65.7} + \dfrac{1}{2} \dfrac{F}{1516} \times \dfrac{F \times 0.1}{65.7} = 1$

gives $F = \underline{418 \text{ kN}}$

Example 6.4

What is the capacity of the column of Example 6.3 for buckling about the major axis, assuming that the loading acts at an effective eccentricity of 100 mm from the column faces such as to induce double curvature bending?

Solution

From section tables, $r_x = 9.98$ cm, $S_x = 652.0$ cm^3

$\quad L/r_x = 3100/89.8 = 34.52$ and noting from *Table 23* that strut curve *b* is appropriate from *Table 24*, $p_c = 256$ N/mm^2

$\therefore$ $\quad P_c = 256 \times 75.8 \times 10^{-1} = 1940.4$ kN

From *Cl. 4.2.5*, $M_{cx} = 275 \times 652.0 = 179.3$ kN m

Total effect eccentricity $= D/2 + 100$ mm $= 204.8$ mm

and since $\beta = -1$ (double curvature) from equation (6.12), $m = 0.43$, giving $mM_x = 0.43 \times F \times 0.205$ kN m

$\therefore$ from equation (6.11)

$$\frac{F}{1940.4} + \frac{F \times 0.205 \times 0.43}{179.3} + \frac{1}{2} \frac{F}{1940.4} \times \frac{F \times 0.205 \times 0.43}{179.3} = 1$$

gives $F = \underline{894 \text{ kN}}$

This second example shows the benefit of using the *m*-factor to allow for the shape of the moment diagram since using $m = 1$ gives $F = 510$ kN, i.e. recognition of the less severe effect on overall buckling strength leads to almost a 40% gain in design capacity.

6.2.2   Case (2): Lateral–torsional buckling

The type of behaviour described above normally occurs only for I- and H-sections bent about their minor axis, for torsionally stiff sections such as tubes, or for strong-axis bending of I- and H-sections when the possibility of out-of-plane deformation is eliminated by the presence of an effective system of alteral bracing. I- or H-sections bent about their major axis normally collapse by buckling in a mode that involves a combination of weak-axis bending and twisting; such behaviour is directly analogous to the lateral–torsional buckling of beams discussed in Chapter 5.

The elastic lateral–torsional buckling of beam columns may be analysed in a manner very similar to the approach adopted for beams [2]. For sections having the normal proportions of columns, manipulation and simplification of the analysis results in an expression for the combination of axial load F and major-axis moment M (assumed for the present to be uniform along the member's length) that is analogous to equation (6.10).

$$\frac{F}{P_{cry}} + \frac{M}{M_E(1 - F/P_{crx})} = 1.0 \qquad (6.13)$$

However, in equation (6.13) P_{cry} is now the critical load for buckling as a strut about the minor axis and M_E is the critical moment for lateral–torsional instability under pure moment. Thus equation (6.13) correctly represents the two extreme cases corresponding to $M = 0$ and $F = 0$. The amplification factor in the second term allows for the enhancement of the applied end moments in the manner shown in Figure 6.6; the importance of this effect depends upon the member's major-axis slenderness, i.e. it is dependent upon P_{crx}.

When the problem is considered as one of the true ultimate strength of the member, analysis and test data show that equation (6.13) provides a reasonable fit to the results providing P_{cry} and M_E are replaced by the strut and beam strengths determined from Section 4.2.2 and equation (5.1) respectively. Since the value of F will be limited to P_{cy}, which must be much smaller than P_{crx}, the effect of the amplification factor may be neglected, leading to the design condition of *Cl. 4.8.3.3.2.*

$$\frac{F}{P_{cy}} + \frac{M}{M_b} = 1.0 \qquad (6.14)$$

Once again the strength of members subjected to unequal end moments M and βM is rather higher. This effect may be approximated closely by replacing M in equation (6.14) with an equivalent value $M = m_{LT}M$ with the value of m_{LT} being obtained from equation (6.12). Since a reduced moment is being used to check overall stability, it will also be necessary to ensure against local overstressing at the more heavily loaded end using (6.4) or (6.3).

Example 6.5

What is the capacity of the column of the previous example for buckling about the minor axis?

Solution
In this case it is first necessary to determine the member's lateral–torsional buckling strength as a beam M_b using the procedures of Chapter 5.
    From *Cl. 4.3.6.7*, $\lambda_{LT} = uv\lambda\sqrt{\beta_w}$ which, using $u = 0.9$, $x = 14.1$, and $\beta_w = 1.0$ from *Cl. 4.3.6.9* according to *Cl. 4.3.7.6*,

$\quad\quad \lambda = 59.7$ from Example 6.3 and $v = 0.852$ from *Table 19*, gives
$\quad\quad \lambda_{LT} = 1.0 \times 0.9 \times 0.852 \times 59.7 \times 1.0 = 45.8$

From *Table 11* corresponding value of $p_b = 248$ N/mm²

$\quad M_b = 248 \times 652 \times 10^{-3} = 162$ kN m

From Example 6.3, $P_{cy} = 1516$ kN

$\quad$ ∴ using equation (6.14), $\dfrac{F}{1516} + \dfrac{F \times 0.20 \times 0.43}{162} = 1$

giving $F = \underline{840\ kN}$ (cf. 894 kN for major-axis buckling)

In this example the ratio $P_{cy}/P_{cx} = 0.8$ and so both checks are necessary. Only two factors contributed to the different values of P_c: the value of λ and the change in the column curve. Other factors which could affect this include end restraint and intermediate bracing that is effective in one plane only, since both of these would lead to the use of different effective lengths in the two planes.

6.2.3   Case (3): Biaxial bending

The most general type of beam-column problem, which automatically incorporates the two previous cases, is the biaxially loaded member of Figure 6.5 (3). Even in the elastic range, analysis of the problem is extremely complex and explicit closed-form solutions cannot be obtained [8, 9]. Thus design equations must be based on an intuitive extension of the procedures of the two previous sections, properly checked against numerical and experimental data [2]. It is therefore convenient to discuss the basis for the design approach of BS 5950: Part 1 in this case from a more qualitative standpoint.
    The main features of the design of a beam column may conveniently be displayed on a three-dimensional interaction diagram of the type shown in Figure 6.10. In this, each of the three axes corresponds to one of the load

components: compression F, major-axis moment M_x or minor-axis moment M_y. A safe design is one which may be represented by a point inside the appropriate failure surface. Because the exact form of the inter-action varies with the slenderness of the member, the shape of this surface will be a function of a member's slenderness, with very stocky members being associated with a convex interaction of the type already illustrated in Figure 6.1 for $L/r \to 0$. When one load component is absent the 3-D surface becomes a 2-D plane, for example when only F and M_x are present a safe design is one that plots below the curve joining the end points on the F and M_x axes appropriate to the member's slenderness.

For the full biaxial problem of Figure 6.5 (3), *Cl. 4.8.3.3.2* gives the design condition as

$$\frac{m_x M_x}{M_{ax}} + \frac{m_y M_y}{M_{ay}} \leq 1 \tag{6.15a}$$

$$\frac{m_{LT} M_x}{M_{ab}} + \frac{m_y M_y}{M_{ay}} \leq 1 \tag{6.15b}$$

in which $M_{ab} = M_b(1 - F/P_{cy})$ \hfill (6.16a)

$$M_{ax} = \frac{M_{cx}(1 - F_c/P_{cx})}{1 + \frac{1}{2}F_c/P_{cx}} \tag{6.16b}$$

$$M_{ay} = \frac{M_{cy}(1 - F_c/P_{cy})}{1 + \frac{1}{2}F_c/P_{cy}} \tag{6.16c}$$

Inequality (6.15) therefore locates a point in the M_x, M_y plane of Figure 6.10, the end points of the curve defining this point having previously been deter-mined by separate consideration of the F, M_x and F, M_y interactions. For sim-plicity the M_x, M_y interaction is taken as linear although some evidence exists to suggest that this is actually convex. In determining the quantities M_{ax} and M_{ay} the procedures reflect the different possible modes of failure illustrated in Figure 6.5 (1) and 6.5 (2) and described in Sections 6.2.1 and 6.2.2.

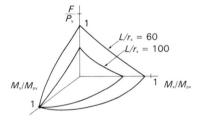

Figure 6.10   Interaction surface for slender beam columns.

The second check in equation (6.16) governs failure in the plane of the applied moments (Figure 6.5 (1)), while the first controls out-of-plane buckling (Figure 6.5 (2)). In the most general case both must be checked since it will not be known in advance which will govern. However, when $P_{cx} > 1.5 P_{cy}$ the second condition will normally govern. Assuming equal degrees of end fixity in both planes (so that $l_x \simeq l_y$), the value of P_{cx} for 'normal sections' will often be found to exceed comfortably that of P_{cy}, with the result that equation (6.16(b)) will more often control.

An alternative, simpler but more conservative expression, which is analogous to (6.3), is also permitted. This is

$$\frac{F_c}{P_c} + \frac{m_x M_{ax}}{p_y Z_x} + \frac{m_y M_y}{p_y Z_y} \le 1 \qquad (6.18)$$

$$\frac{F_c}{P_{cy}} + \frac{m_{LT} M_x}{M_b} + \frac{m_y M_y}{p_y Z_y} \le 1$$

in which P_c = compression resistance considering the possibility of buckling about either axis.

Because (6.15) and (6.18) use equivalent moments mM, a separate check against exceeding the local capacity of the member at its most heavily loaded cross-section is also necessary. This is achieved by using either (6.4) or (6.3) in the form of a pure strength check, i.e. M_{cx} should always be the pure cross-sectional bending capacity.

Example 6.6

Check the ability of a 3.1 m long 203 × 203 × UC 60 of S275 steel to carry a compressive load of 340 kN, assuming that this acts at effective eccentricities of 100 mm from the column face such as to produce single curvature bending about the $y - y$ axis and double curvature bending about the $x - x$ axis.

Solution
P_{cy} From section tables, $A = 75.8$ cm^2, $r_y = 5.19$ cm
   Take $p_y = 175$ N/mm^2
   $\lambda = 3100/519 = 59.7$
   Corresponding p_c from *Table 24* = 200 N/mm^2
   $\therefore P_{cy} = 200 \times 7580 \times 10^{-3} = \underline{1516 \text{ kN}}$

P_{cx} From section tables, $r_x = 8.98$ cm
   $\lambda = 3100/8.98 = 34.52$
   Corresponding p_c from *Table 24* = 256 N/mm^2
   $\therefore P_{cx} = 256 \times 7850 \times 10^{-3} = \underline{1940.4 \text{ kN}}$

M_{cy} From section tables, $S_y = 302.8 \text{ cm}^3$
$\quad M_{cy} = 275 \times 302\,800 \times 10^{-6} = \underline{83.27 \text{ kN m}}$

M_{cx} From section tables, $S_x = 652.0 \text{ cm}^3$
$\quad M_{cx} = 175 \times 652\,000 \times 10^{-6} = \underline{179.3 \text{ kN m}}$

M_b Using *Cl. 4.3.6.8*, $u = 0.9$, $x = 14.1$
$\quad \lambda/x = 59.7/14.1 = 4.23$
From *Table 19*, corresponding $v = 0.852$
$\lambda_{LT} = 1.0 \times 0.9 \times 0.852 \times 59.7 = 45.8$
From *Table 20*, corresponding $p_b = 248 \text{ N/mm}^2$.
$M_b = 248 \times 652\,000 \times 10^{-6} = \underline{162 \text{ kN m}}$

M_{ax} From equation (6.16b), $M_{ax} = 179.3 \times \dfrac{1 - 340/1940.4}{1 + \frac{1}{2} \times 340/1940.4}$

$$= 136 \text{ kN m}$$

From equation (6.16a), $M_{ax} = 162\,(1 - 340/1516) = \underline{126 \text{ kN m}}$

M_{ay} From equation (6.16c), $M_{ay} = 83.27 \times \dfrac{1 - 340/1516}{1 + \frac{1}{2} \times 340/1516}$

$$= 58.2 \text{ kN m}$$

$m_x M_x$ Since $\beta = -1$ from equation (6.12), $m = 0.43$
$\quad m_x M_x = 340\,(100 + \frac{1}{2} \times 209.6) \times 0.43$
$\qquad = \underline{29.9 \text{ kN m}}$
Value of $m_{LT} M_x = \underline{29.9 \text{ kN m}}$

$m_y M_y = 340 \times 100 \times 34.0 \text{ kN m}$ $(m = 1.0)$

$\quad$ Using (6.15a) $\dfrac{29.9}{136} + \dfrac{34.0}{58.2} = 0.220 + 0.584$
$$= \underline{0.804}$$

$\quad$ Using (6.15b) $\dfrac{29.9}{126} + \dfrac{34.0}{58.2} = 0.237 + 0.584$
$$= \underline{0.821}$$

Both checks are satisfactory

Check local strength at most heavily loaded cross-section; this will be at either end where the loads are $F = 340 \text{ kN}$, $M_x = 69.6 \text{ kN m}$ and $M_y = 34.0 \text{ kN m}$.

Using equation (6.5), $n = 340/(275 \times 75.8 \times 10^{-1}) = 0.163$
Change formula for S_{rx} at $n = 9.4\,(209.6 - 2 \times 14.2)/7640$
$$= 0.223$$

$$\therefore S_{rx} = 656\,000 - \left[\frac{7640^2}{4 \times 14.2}\right]0.163^2$$

$$= \underline{656\ cm^3}$$

and $M_{rx} = 275 \times 656 = \underline{180\ kN\ m}$

Change formula for S_{ry} at $n = 14.2 \times 209.6/7640$

$$= 0.390$$

$$\therefore S_{ry} = 305\,000 - \left[\frac{7640^2}{4 \times 209.6}\right]0.163^2$$

$$= \underline{303\ cm^3}$$

and $M_{ry} = 275 \times 303 = \underline{83\ kN\ m}$

$$\left(\frac{69.6}{180}\right)^2 + \left(\frac{34.0}{83}\right) = 0.150 + 0.410$$

$$= \underline{0.560} \qquad \text{Satisfactory}$$

Using the simpler alternatives of equations (6.18) and (6.3)

$$\frac{340}{1516} + \frac{29.9}{275 \times 584} + \frac{34.0}{275 \times 201} = 0.224 + 0.186 \times 0.615$$

$$= 1.025$$

$$\frac{340}{1516} + \frac{29.9}{162} + \frac{34.0}{275 \times 201} = 0.224 + 0.185 \times 0.615$$

$$= 1.024$$

and section is unsafe for overall buckling

$$\frac{340}{75.8 \times 275 \times 10^{-1}} + \frac{69.6}{179.3} + \frac{34.0}{83.3} = 0.163 + 0.388 + 0.408$$

$$= \underline{0.959} \qquad \text{Satisfactory}$$

This biaxial example does, of course, incorporate all of the component problems covered in the earlier examples. In practice, where the requirement is usually one of checking the adequacy of a trial section rather than one of determining the precise load-carrying capacity, use of the simpler inequalities (6.18) and (6.3) will generally prove easier. However, if the trial section just fails to meet these requirements (as is the case in this example), then recourse to the more exact provisions of (6.15) and (6.4) may well enable that section to be used.

Exercises

1   Use (6.3) to determine the major-axis moment that can safely be carried by a 254 × 254 UC 89 in S275 steel that is already subjected to a tension of 1450 kN.

[166 kN m]

2   Compare the answer to Exercise 1 with the result obtained using (6.4) in conjunction with the formulae of the *Structural Steelwork Handbook*.

[209 kN m]

3   A 203 × 203 UC 52 is subject to an axial tension of 1125 kN. Assuming S275 steel, can it also withstand moments of 53 kN m and 14 kN m about its major and minor axis respectively?

[Yes, assuming (6.4)]

4   Determine the compressive load that can be carried by a 406 × 178 UB 60 in S275 steel over a height of 5.6 m, assuming that it is braced against out-of-plane failure and that the maximum moment about its major axis is 72 kN m.

[1380 kN]

5   Determine the load-carrying capacity of a 305 × 305 UC 118 of effective height 3.6 m in S275 steel, assuming it to be an external column in a simply connected frame with beam reactions of 105 kN.

[3200 kN]

6   Determine the end moments that can safely be carried by a 305 × 127 UB 37 in S275 steel, assuming that these are in the ratio 1 : 0.3 and that they produce bending about the section's major axis. Take the member length as 4.8 m and allow for the presence of a 205 kN compressive load.

[10 kN m]

7   Check the ability of a 203 × 203 UC 60 of height 3.8 m in S275 steel to carry the following load combination:

F(compressive) 750 kN
M_x 52 kN m   (top)   0.0 kN m   (bottom)
M_y 13.8 kN m   (top)   11.0 kN m   (bottom)

[Satisfactory]

References

1   Horne, M.R. (1979) *Plastic Theory of Structures*, 2nd edn, Pergamon Press, Oxford.
2   Chen, W.F. and Atsuta, T. (1976, 1977) *Theory of Beam-Columns*, Vols. 1 and 2, McGraw-Hill, New York.
3   Steel Construction Institute (1997) *Steelwork Design Guide to BS 5950: Part 1: 1985. Volume 1 Section Properties, Member Capacities*, 4th edn.
4   Morris, L.J. and Randall, A.L. (1975) *Plastic Design*, Constrado, London.

5 Neal, B.G. (1963) *Plastic Methods of Structural Analysis*, 2nd edn, Chapman and Hall, London.
6 Trahair, N.S. and Bradford, M.A. (1998) *The Behaviour and Design of Steel Structures to AS 4100*, 3rd edn, E & FN Spon, London.
7 Galambos, T.V. (1998) *Guide to Stability Design Criteria for Metal Structures*, 4th edn, Wiley, New York.
8 Culver, G.C. (1966) Exact solution of the biaxial bending equations, ASCE, *J. of Structural Division*, **92**(ST2), 63–83.
9 Culver, G.C. (1966), Initial imperfections in biaxial bending, ASCE, *J. of Structural Division*, **92**(ST3), 119–35.

Chapter 7

Joints – basic concepts

Previous chapters have dealt with the design of different types of member such as beams and columns, with little consideration of the ways in which these are attached to one another to form a structure. However, many fabricators would argue that the economics of a steel structure are much more dependent upon the types of joint used than upon the sizes of the members. The basis for this lies in the fact that typical material costs represent only about 25–50% of the overall cost for the steelwork. It is thus not uncommon for the fabricator's own design staff to suggest modifications to joint details; providing the integrity of the structure is retained this is normally acceptable, since they are better placed to appreciate the equipment available and the effects on cost of various alternatives. Indeed, in some cases joint design is left entirely to the fabricator with the designer of the main structure supplying details of the loads which each connection must transmit together with any particular requirements, for example to provide adequate lateral restraint to the end of a beam.

Connections may involve the use of bolts (of which there are several different types) or welds, or a combination of both; rivets, although they are found in older structures such as railway bridges, are rarely used nowadays. Either type may be used for connections made in the fabricating shop; site connections will usually be bolted. Although it is possible to weld on site, the process is expensive since it requires special staging to provide a working platform, protection from the weather is necessary, the welds must be inspected and problems of access may arise since welding is much easier in certain positions, e.g. from above (downhand).

7.1 Methods of making connections

7.1.1 Bolts

Many different types of structural bolt are available in the UK. Apart from the obvious variations in diameter and length, grade of steel, head type and thread arrangements may differ. *Sections 3.2.1* and *3.2.2* refer to the principal types and give details of the relevant product standards.

Appropriate matching combinations (of bolts, nuts and washers) are specified in *BS 5950: Part 2*.

Bolts are normally used in clearance holes 2 mm (or in the case of M24 and larger 3 mm) greater in diameter than the nominal bolt size. Although bolts are available in Grades 4.6, 8.8, 10.9 and 12.9, the great majority of structural connections are made using 8.8s, with 4.6s generally being reserved for secondary connections, e.g. purlin cleats. The most popular size is M20. (The first number is one tenth of the minimum UTS in kilograms per square millimetre, while the product of both numbers gives the minimum yield stress in kilograms per square millimetre. For example Grade 8.8, min. UTS $= 10 \times 8 = 80$ kg/mm^2, min. yield stress $= 8 \times 8 = 64$ kg/mm^2.)

High Strength Friction Grip or HSFG bolts are made from high-tensile steel and are tightened sufficiently with special torque wrenches to produce a predetermined shank tension, thereby enabling additional shear resistance to develop between the connected plates as a result of friction. Installation is therefore a more critical operation and BS 4604 covers this as well as providing details of suitable design procedures. HSFG bolts may be recognized by their larger diameter head and the additional identifying marks provided. Because of the more onerous installation requirements they should only be used in situations where a genuine need exists, e.g. the shear capacity of Grade 8.8 bolts is insufficient, no movement may be tolerated in the connection when under load, fluctuating loads are present, etc. For ordinary beam to column, beam to beam, splice and column base connections in buildings Gr. 8.8 bolts will normally be suitable. Guidance on particular features of the design of connections using HSFG bolts is available in a CIRIA Technical Note [1].

Although bolts are manufactured in a wide range of diameters and lengths, certain sizes are 'preferred' and are therefore more readily available. Table 7.1 lists these for both 4.6s and 8.8s. Only the fully threaded

Table 7.1   Recommended sizes – fully threaded bolts

Grade	Diameter	Length (mm)		
*4.6	M12	25	–	–
8.8	M16	30	45	–
8.8	M20	45	60	–
8.8	M24	70	85	100

* Intended for use in fixing cold rolled purlins and rails
Fully threaded fasteners should be specified as follows:
8.8 screws to BS EN 24017 with nuts to BS EN 24032
4.6 screws to BS EN 24018 with nuts to BS EN 24034

type is listed as, increasingly, these are the industry preference. The alternative, in which a short length immediately below the head is left unthreaded, gives greater strengths in certain applications but results in the use of more bolt lengths being required on a job. It has been demonstrated that some 90% of the primary connections in a typical multistorey steel frame building can be made using M20 × 60 mm long grade 8.8 fully threaded bolts.

Readers requiring further information on bolts are referred to Chapter 2 of the Steel Designers' Manual [2] or to Chapter 3 of the book by Owens and Cheal [3].

7.1.2   Welds

A weld is produced by passing a high current, typically between 50 and 400 amperes, through an electrode or filler wire so as to produce an arc which completes the path from the power source through the specimen to earth. Sufficient heat is produced – temperatures reached in the arc range between 5000 °F and 30 000 °F (2800–16 700 °C) – to melt both the electrode and the parent metal so that the plates being welded fuse together on cooling. Typical specimens cut from welds are shown in Figure 7.1. Possible embrittlement of the welded area is avoided by ensuring that while hot it is surrounded by an inert gas. This is provided by means of a substance called flux, either directly from the electrode as a core or coating, or, when bare wire is being used, in powder form

Although welded joints produce cleaner lines, thereby avoiding possible corrosion traps, they generally require tighter tolerances than equivalent bolted joints. Also, the reduced preparation and handling must be set

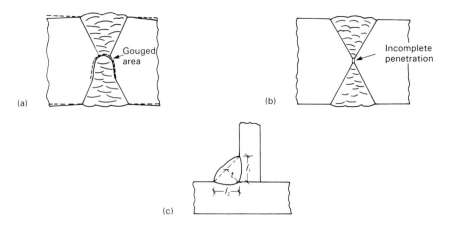

Figure 7.1   Cross-sections of the main types of structural welds. t = throat thickness; l_1 = vertical leg length; l_2 = horizontal leg length. (*After reference 4.*)

against the costs of the skill labour required for the fabrication and subsequent inspection. Because of the obvious difficulty in checking the adequacy of a weld simply by visual means, inspection using more sophisticated methods, including X-ray, magnetic particle inspection (MPI), and ultrasonics, is normally employed. Full details of these techniques are provided in the appropriate British Standards; these are listed by Pratt [4]. In certain cases destructive tests on sample welds may be necessary as specified in BS 709.

Several different welding processes are available for the fabrication of structural steelwork. Probably the most widely used is the manual metal arc process (MMA); others include various automatic and semi-automatic processes such as CO_2, submerged arc and, where large deposition is required, electroslag. Full descriptions of these, together with guidance on the selection of the best process for a particular application, are available in Section 7 of [4] and Chapter 2 of [2].

Figure 7.2 illustrates the two types of weld in common use for structural steelwork. For butt welds the weld metal is placed between the edges of the plates, whereas for fillet welds the weld metal is located on the faces of the plates. various details, namely arrangements of the welds and corresponding edge preparations of the plates, are possible, especially when large welds are required. BS 5135 provides details of these as well as listing the agreed symbols used to specify them on drawings. In certain cases where automatic or semi-automatic processes are to be used some modification may be permitted, providing all parties are agreeable; such agreement is usually based on procedural trials.

Butt welds may be either 'full penetration' or 'partial penetration' as shown in Figure 7.2. The latter type are useful where access from both sides

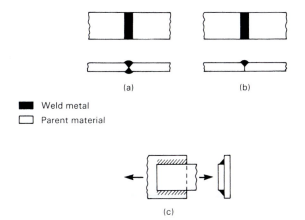

Weld metal
Parent material

Figure 7.2   Basic types of weld: (a) full penetration; (b) partial penetration; (c) fillet weld [4].

is impractical, although this does, of course, result in some eccentricity in the weld area which should be properly allowed for in design. Full penetration butt welds are designed on the basis of equivalence to the parent plate using the design strength of the parent metal, whereas partial penetration butt welds are assumed to possess an area corresponding to the depth of penetration only as explained in *Cl. 6.9.2*. Although full-penetration butt welds are structurally the most efficient (because they enable the full strength of the original cross-section to be utilized), the amount of fabrication involved even for the most usual type of double-V edge preparation tends to make them expensive. They should therefore be used only when circumstances really warrant it. For partial-penetration, single-V butt welds, the efficiency as defined by the ratio of the axial stress in the plates to the maximum stress in the weld (allowing for bending effects) varies between about 20 and 60%, as plate thickness increases from 10 to 40 mm.

The load-carrying capacity of a fillet weld is obtained as the product of the throat area and the design strength of the weld p_w as given in *Table 37*. Figure 7.3 shows how for a 90° fillet weld the effective throat size is determined as the dimension '*a*' subject to an upper limit of 70% of the effective leg length.

The values of p_w provided are based on experimental data [5] and correspond to 0.47 × UTS of the weld metal. Weld groups subject to a complex stress system should be designed using a 'vector sum' approach as indicated by *Cl. 6.8.7.3* such that the resultant stress does not exceed p_w. Useful comments on the implementation of this approach are available in Chapter 4 of [3] and Chapter 24 of [2].

This requires the force per unit length transmitted by the weld to be calculated using the elastic section properties of the weld group and the set of applied forces and moments.

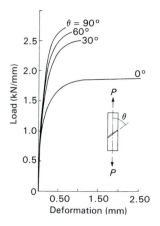

Figure 7.3   Strength and ductility of fillet welds as a function of load orientation [5].

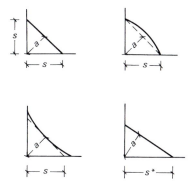

Figure 7.4   Definition of throat sizes for fillet welds.

Alternatively, account may be taken of the directional aspect of weld strength [5] as illustrated by the test data shown in Figure 7.4. This indicates that transversely loaded welds ($\theta = 90°$) are significantly stronger than longitudinally loaded welds ($\theta = 0°$) – although this additional strength is obtained at the expense of ductility. *Cl. 6.8.7.3* permits this to be recognized by resolving the force per unit length in the weld into longitudinal F_L and F_T components, see Figure 7.5. These components should then be checked against the respective capacities P_L and P_T using:

$$P_L = p_w a$$

$$P_T = K p_w a$$

in which $K = 1.25 \left[\dfrac{1.5}{1 + \cos^2 \theta} \right]^{\frac{1}{2}}$

and θ is defined in Figure 7.5.

Throughout the weld's length, the following must be satisfied:

$$\left(\frac{F_L}{P_L} \right)^2 + \left(\frac{F_T}{P_T} \right)^2 \leq 1$$

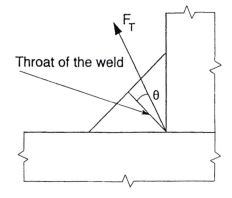

Figure 7.5   Force components in weld.

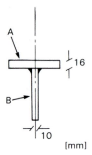

Figure 7.6

Example 7.1

Two plates are connected by means of a pair of fillet welds as shown in Figure 7.6. Assuming material and electrodes to *Cl. 3.2.3* of BS 5950: Part 1, what size welds are required in order that a tensile force equal to the full strength of plate B can be developed?

Solution
From *Table 9*, for 16 mm material p_y $= 275 \text{ N/mm}^2$
∴ tensile strength of plate B per unit width $= 10 \times 1 \times 275$
$$= 2.75 \text{ kN/mm}$$

From *Cl. 6.8.7*, weld strength $= 2ap_w$
which, taking p_w as 220 N/mm² from *Table 37* $= 2a \times 220$
$$\therefore 2a \times 220 = 2750$$
$$a = 6.25 \text{ mm}$$
and effective throat size of each weld $= 8$ mm.

In this case the convention that weld sizes follow a pattern has been followed; thus since 6 mm welds are fractionally too small, the next standard size of 8 mm has been specified.

7.2   General principles of connection design

Structural connections are required when two different members must be joined together, for example a beam-to-column connection, or when an individual member is too large for complete shop fabrication, for example splices are normally provided at about every other floor level in multi-storey frames. Table 7.2 illustrates one example of each of the main types of steelwork connection.

Curved trusses for the roof of Baltic Quay.

Whatever the form of connection used certain general design principles should be observed.

1 Connections subject to impact or vibration or load reversal (other than that due solely to wind action) should not use bolts in clearance holes.
2 The use of very large diameter bolts (greater than about M30), especially HSFG bolts, can lead to problems with installation; a better design will usually result if a larger number of smaller bolts are used.
3 Standardize on one size and grade of bolt in a connection and limit as far as possible the number of different sizes and grades in the structure. Where different grades are required the inadvertent use of a lower-grade bolt in place of the specified higher grade may be avoided by adopting different sizes, for example M20 Grade 8.8 bolts, M24 HSFG bolts.
4 Before specifying HSFG bolts check for possible problems with installation and inspection. Use only when necessary.
5 For welded joints subject to fatigue, as in crane rails, check Part 10 of BS 5400; try to avoid the lower class detail, i.e. those with poor fatigue performance.
6 Do not specify larger fillet welds than are necessary. Avoid butt welds,

which require expensive preparation, if fillet welds of a reasonable size would suffice.

7 Consider the number of workshop operations required, for example for a member of all-welded construction apart from one end connection that requires drilling the cost of the separate process will be excessive.

8 Avoid connection plates which require a large number of cuts. For trusses the gusset plates may be omitted altogether in certain circumstances and the joints made directly to the member.

9 Avoid unnecessary splices in columns; unless the potential material savings are large or special factors are present, it will often be cheaper to run the heavier section through.

10 The use of bolts and welds to resist the same load component in a connection is permissible only if HSFG bolts are used and the bolts are fully torqued after the welds are made. (With ordinary bolts slip would result in all of the load being transferred to the welds.)

7.3 Modes of failure for fasteners

7.3.1 Bolts

Inspection of the example connections of Table 7.2 shows that the actual loading on the bolts will be either shear, tension or a combination of the two. For most forms of simple connection it is customary to design the bolts for shear only. The basic connection problem is therefore as shown in Figure 7.7 in which several bolts in line are each subjected to a shearing action at the plate interface. Except in the case of long joints, defined by *Cl. 6.3.2.5* as exceeding 500 mm in length, the load on each bolt may be assumed equal.

This leads to the four possible types of failure shown in Figure 7.8, only one of which actually depends upon bolt strength. The first of these – tearing at the net section of either plate – has already been covered in Chapter 3. Shearing of the plate beyond the end fastener should not occur providing the end distance exceeds $1.25d$ or $1.40d$ as defined by *Cl. 6.2.2*. This leaves the two most important failure modes: shearing of the bolt

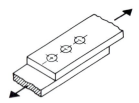

Figure 7.7   Bolts in shear [6].

Table 7.2 Examples of the main forms of steelwork connection (*Bates, Constrado Publications.*)

Type	Use
1	Beam to beam
2	Beam to column (transmits shear only)
3	Beam to column (full moment connection)
4	Truss connection

Gusset plate

Table 7.2 continued

Type	Use
5	Column baseplate

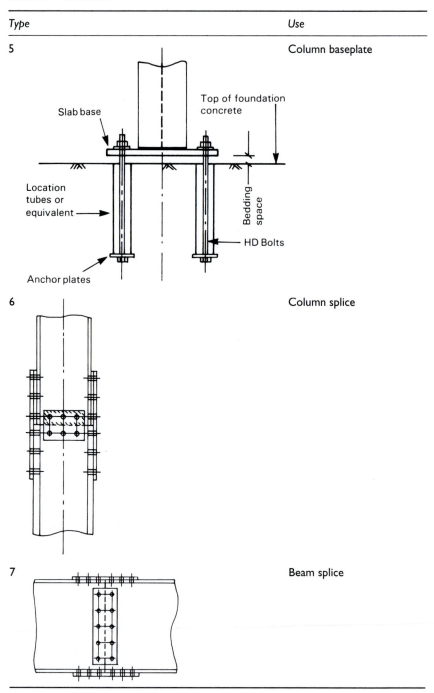

6 — Column splice

7 — Beam splice

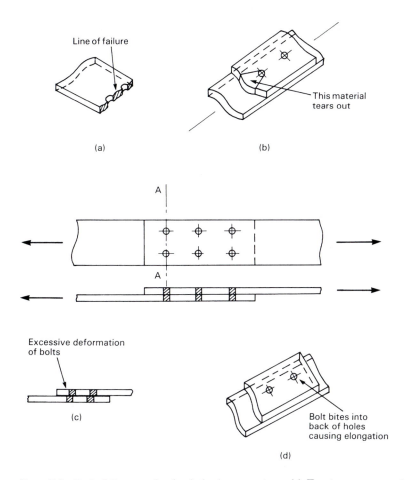

Figure 7.8   Basic failure modes for bolted connections. (a) Tension at net section AA; (b) end failure of plate; (c) shear of bolts; (d) bearing.

itself or bearing of the plate immediately behind the bolt, as illustrated in Figures 7.8(c) and 7.8(d) respectively.

In determining the shear capacity of a bolt it is important to distinguish between the two cases:

1   at least one shear plane passes through the threaded portion;
2   threads do not occur in the shear plane.

Although case (1) is much more common, higher strengths can be developed for case (2) and this is recognized by permitting the use of the shank area A in such cases. For case (1) the area available to resist shear

A_s will be the tensile area A_t. Thus the shear capacity P_s of one bolt in a condition of single shear as illustrated in Figure 7.8 is given by *Cl. 6.3.2* as

$$P_s = p_s A_s \tag{7.1}$$

in which p_s = shear strength obtained from *Table 30*.
and A_s = shear area obtained from *Cl. 6.3.1*.

The values given for p_s are the lesser of 0.69 times the yield strength or 0.48 times the ultimate strength of the fastener.

Bolts passing through more than two plates will possess a higher shear capacity since the total shear will be divided between the interfaces. As an example, the bolts in the web cover plates of the beam splice shown in Table 7.2 will be in double shear and the appropriate shear area for use in equation (7.1) will therefore be $2A_s$. Figure 7.8 illustrates the two cases. Caution is necessary, however, if the bolts pass through a total thickness of material (including packing) significantly in excess of the bolt diameter as bending of the bolt will reduce the available shear capacity as explained in *Cl. 6.3.2.2*.

Bearing failure occurs when the bolt bites into the rear edge of its hole causing elongation and eventual tearing. Unless a low-strength bolt is used with higher-strength plates then the governing factor will be the bearing strength of the weakest connected ply, given by *Cl. 6.3.3.3*

$$P_{bs} = K_{bs}\, dt p_{bs} \tag{7.2}$$

in which d = effective, i.e. nominal, diameter of the bolt
t = thickness of connected ply
p_{bs} = bearing strength of the connected parts obtained from *Table 31*.

K_{bs} is a factor dependent on the type of hole; for standard clearance holes $K_{bs} = 1.0$

Equation (7.2) assumes that sufficient material is present between the back face of the hole and the end of the plate. If this is less than twice the bolt diameter then bearing capacity must be reduced *pro rata* by noting that P_{bs} should not succeed $0.5 K_{bs} e t_p P_{bs}$.

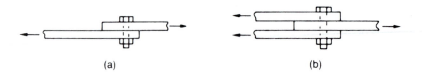

(a)   (b)

Figure 7.9   Bolts in shear: (a) single shear; (b) double shear.

The values given for p_{bs} are based on considerations of serviceability, since actual bearing failure of the plate occurs at such high stresses that deformations will have become unacceptably large at a much earlier stage. For bolts in clearance holes the figure of 0.65 (ultimate strength + yield strength) used in *Table 31* reflects the approximate dependence of a suitable figure on the mean of the ultimate tensile stress and the yield stress.

For any given situation the strength of a bolt will clearly be the lesser of its capacities in shear and bearing. Bearing in the thinner plate will normally control for plate thicknesses up to about one half the bolt diameter.

The reason that bearing will not normally be critical is the extremely high values of p_{bs} given in *Table 31*

In the same way that fasteners should not be placed too near the ends of the connected plate they must also be suitably spaced both from each other and from the edges of the plates. The rules given in *Cl. 6.2* are based on several practical considerations. These include the provision of sufficient space between bolts to permit proper tightening, limiting the distance between bolts in compressive regions, both to avoid buckling and to avoid corrosion by ensuring adequate bridging of the paint film between plates [7].

Example 7.2

Calculate the strength of the bolts in the lap splice in Figure 7.10 assuming the use of M20 Grade 8.8 bolts in 22 mm clearance holes and S275 plate.

Solution
Bolts are in single shear, from equation (7.1) shear capacity per bolt

$$= 375 \times 245 \text{ N} = 91.9 \text{ kN}$$

Bearing capacity of thinner plate per bolt, from equation (7.2)

$$= 450 \times 20 \times 16 \text{ N} = 144 \text{ kN}$$

The full value is appropriate since the end distance $e \not< 2d$.

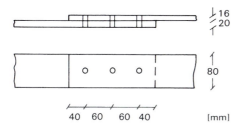

Figure 7.10

Clearly capacity is controlled by strength in shear. Therefore joint capacity in tension as governed by bolt strength = 3 × 91.9 = 276 kN.

Had bearing governed it could only have been improved by increasing either plate strength or plate thickness.

Moment resisting beam-to-column connections often contain regions in which the bolts will be required to transfer load by direct tension, such as the upper bolts in the end-plate connection shown in Table 7.2. The capacity P_t of such bolts is determined from the equivalent of equation (7.1) with tensile area A_t as specified in BS 3643 and tensile strength as given in *Table 34*. One rather contentious issue in the design of such connections concerns the additional forces induced in the bolts as a result of so-called 'prying action'. If one of the connected plates is sufficiently flexible to deform appreciably as illustrated in Figure 7.11, then some allowance for the resulting bending of the bolts would appear to be in order. One suggestion [6] is that the nominal bolt forces be scaled up by the factor.

$$\left(\frac{3b}{8a} - \frac{t^3}{20} \right) \tag{7.3}$$

However, *Cl. 6.3.4.2* allows prying to be treated in a simpler way by requiring that the bolt tensile strengths of *Table 34* be reduced by 20%. As an alternative *Cl. 6.3.4.3* permits the full strength to be used providing one or more of 4 conditions is met. Two of these define situations in which prying will not be present, whilst the others explain how the prying contribution should be calculated and then included in the total applied tension in the bolt.

Where both shear and tension are present in the bolts, as with the upper bolts in the bracket connection of Table 7.2, then their combined effect may conveniently be assessed from a suitable interaction diagram.

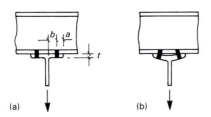

Figure 7.11   Prying action causing bending of tension bolts passing through a flexible flange. (a) Rigid flange; (b) flexible flange. (*Kulak, Adams and Gilmor, 1990.*)

Cl. 6.3.4.4 of BS 5950: Part 1 specifies a trilinear diagram of the type shown in Figure 7.11; this is represented by

$$\frac{F_s}{P_s} + \frac{F_t}{P_t} \ngtr 1.4 \qquad (7.4)$$

in which F_s and F_t are the applied shear and tension and P_s and P_t are the shear and tension capacities.

Although the experimental data in Figure 7.12 are for the higher grades of bolt used in the USA [6], recent tests on M20 Grade 4.6 bolts [7] support this general shape of interaction as shown in Figure 7.13.

Example 7.3

The tee-stub shown in Figure 7.14 is part of a beam-to-column connection which is required to transfer 350 kN in tension and 110 kN in shear. Check whether four M20 Grade 8.8 bolts will be adequate.

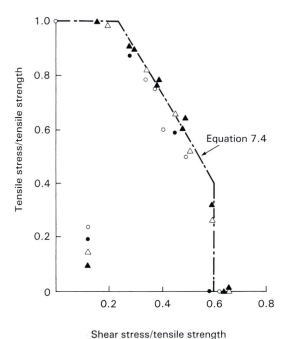

Figure 7.12 Trilinear interaction curve for bolts under combined tension and shear comparison with test data. (*From reference 6.*)

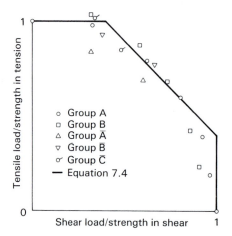

Figure 7.13   Comparison of test results of reference [7] for M20 grade 4.6 bolts in shear and tension with equation (7.4), evaluation based on experimentally obtained values of P_s and P_t.

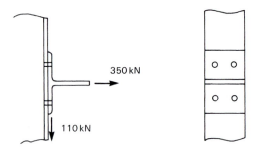

Figure 7.14

Solution

Tensile load per bolt $F_t = 350/4 = 87.5$ kN

Shear load per bolt $F_s = 110/4 = 27.5$ kN

From equation (7.1), using $A_s = A_t$ (assumes shear plane passes through threads)

$$P_s = 375 \times 245 \text{ n} = 91.9 \text{ kN}$$
$$P_t = 450 \times 245 \text{ N} = 110.3 \text{ kN}$$

Check inequality (7.4)

$$27.5/91.9 + 87.5/110.3 = 0.30 + 0.79$$
$$= 1.09 \qquad \text{Satisfactory}$$

7.3.2   Bolted connections using HSFG bolts

Figure 7.15 illustrates the type of load-deflection curve obtained from a typical test on an HSFG bolted connection loaded in shear. Of particular importance is the plateau corresponding to the load level at which slip between plies occurs, since this is absent for normal shear/bearing type connections. In BS 5950: Part 1, ordinary parallel shank friction grip fasteners are designed on the serviceability condition of slip presented as an ultimate check. In those cases where such connections will have slipped into bearing at some stage between working and ultimate load a bearing capacity check is also necessary. For waisted-shank fasteners BS 5950: Part 1 regards slip as 'failure'; since such connections must be designed on a non-slip basis the bearing check is unnecessary.

The slip resistance of parallel shank fasteners P_{sL}, is given by *Cl. 6.4.2* as

$$P_{sL} = 1.1 \, K_s \mu P_0 \tag{7.5}$$

in which P_0 = minimum shank tension from BS 4604
$\quad\quad \mu$ = slip factor $\not> 0.55$
$\quad\quad K_s$ = 1.0 for fasteners in clearance holes (lower values are necessary in the case of oversize or slotted holes)

Slip factors are specified in *Table 35*. Since these are safe 'average' values, better results may be achieved (if the situation warrants the additional effort) by conducting specific slip tests as specified in BS 4604. For clean, shot blaster surfaces a figure of 0.5 is usually appropriate; if the surface has been cleaned simply by wire brushing, this should be reduced to 0.3.

Bearing is checked using equation (7.2) with K_{bs} being taken as 1.5 to recognize the relaxation of any requirements on acceptable deformation, i.e. the check is on strength only since the bolts will actually be in bearing only once the serviceability limit has been passed. For end distances of less than $3d$ a *pro rata* reduction is required.

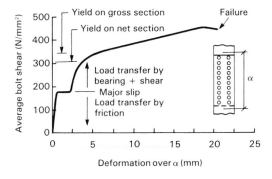

Figure 7.15   Overall behaviour of a friction type connection showing effect of slip [9].

Example 7.4

Repeat Example 7.2 assuming the use of M20 parallel-shank HSFG bolts in clearance holes.

Solution
Take $\mu = 0.50$ and obtain $P_0 = 144$ kN from BS 4604, note that the joint has only one pair of surfaces in contact.
From *Cl. 6.4.2*, slip resistance per bolt $= 1.1 \times 1.0 \times 0.45 \times 144$
$$= 79.2 \text{ kN}$$
From *Cl. 6.4.4*, bearing resistance per bolt $= 1.5 \times 20 \times 16 \times 460 = 221$ kN
This assumes an end distance $e \nleq 3d$. However, since $e = 40$ mm, bearing resistance should be reduced to $0.5 \times 40 \times 16 \times 460 = 147$ kN.
Clearly capacity is controlled by slip resistance of the bolts and tensile capacity of connection as governed by fastener strength
$= 3 \times 79.2 = \underline{237.6 \text{ kN}}$ Satisfactory

In the case of waisted-shank fasteners, equation (7.5) is appropriate for checking slip, providing the factor 1.1 is replaced by 0.9. HSFG bolts in tension may be designed for $0.9P_0$, while combined shear and tension is controlled by the linear interaction of *Cl. 6.4.5*.

Example 7.5

Repeat Example 7.3 assuming the use of M20 parallel-shank HSFG bolts in clearance holes.

Solution
Taking $\mu = 0.50$ and $P_0 = 144$ kN from BS 4604, from *Cl. 6.4.2*.
slip resistance per bolt $= 1.1 \times 1.0 \times 0.50 \times 144 = 79.2$ kN
Tensile load per bolt $= 350/4 = 87.5$ kN
Shear load per bolt $= 110/4 = 27.5$ kN
Check interaction equation of *Cl. 6.4.5*.

$$\frac{27.5}{79.2} + \frac{87.5}{0.9 \times 144} = 0.35 + 0.68$$
$$= \underline{1.07} \qquad \text{Not satisfactory}$$

However if the connection is only required to function as non-slip under service loads, then the second term becomes 0.55 and the arrangement is satisfactory.
    Although lack of fit between the connected plates may affect the degree

of preload that may be achieved in each bolt in a connection, experimental evidence [8] suggests that this will not necessarily impair the subsequent performance of that connection.

Tightening of HSFG bolts to produce a given preload is normally controlled by one of the following:

1   *Torque control.* Use a calibrated manual wrench or a power wrench set to cut out at a given torque; the value of torque used must be related to the required preload.
2   *Turn of nut.* After preliminary tightening with an ordinary podger spanner sufficient to bring the surfaces into contact, the nut and bolt shank end are marked, the nut is then turned further relative to the shank – typically one half or three-quarters of a turn is used – to provide a tension which normally exceeds the minimum proof load of the bolt.
3   *Direct tension indication.* A load-indicating washer such as Coronet, or load-indicating bolt (Lib) is used to provide a direct indication of bolt tension; the principle is one of tightening to provide the desired gap under the bolt head, i.e. either device squashes as tension is increased but in a controlled way.

Of the three methods the use of direct indication, although more expensive in that special washers or bolts are required, is now the most widely used [2] in the UK, although turn-of-the-nut is popular in North America.

Tightening of HSFG bolts is discussed from a practical point of view in the paper by Burdekin [9].

Example 7.6

A 150 × 20 mm tie in S275 steel carrying 400 kN requires a splice within its length. Design a suitable arrangement using a single-sided cover plate and (a) bolts in shear, (b) HSFG bolts, (c) fillet welds.

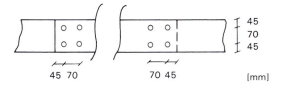

45
70
45

45 70   70 45

[mm]

Figure 7.16

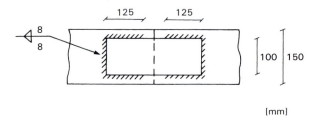

[mm]

Figure 7.17

Solution
(a) *Shear-type bolted connection*
Try M20 Grade 8.8 bolts in 22 mm holes, use 2 rows of bolts.
From *Cl. 6.2.1.1*, minimum spacing $= 2\frac{1}{2} \times 20 = 50$ mm
From *Cl. 6.2.1.2*, maximum spacing $= 14 \times 20 = 280$ mm
From *Cl. 6.2.2.3*, minimum edge and end distance (assuming a sawn edge) $= 1.25 \times 22 = 28$ mm
Try the arrangement shown in Figure 7.16.
From *Cl. 6.3.2*, capacity per bolt in single shear $= 375 \times 245$ N $= \underline{91.9 \text{ kN}}$
From *Cl. 6.3.3.3*, capacity per bolt in bearing in 20 mm plate $= 450 \times 20 \times 20$ N $= \underline{180 \text{ kN}}$
Therefore strength in shear governs and number of bolts required $= 400/91.9 = 4.4$
Use 6 bolts in 2 rows of 3.
Check total lap length against *Cl. 6.3.2.3*.

Lap length $= 2 \times 45 + 2 \times 70 = 230$ mm

No reduction in bolt strength required as this is less than 500 mm.
Capacity of plate at net section, using $p_y = 265$ N/mm² from *Table 9* $= 265 \times (150 - 2 \times 22) \times 20$ N $= \underline{562 \text{ kN}}$ Satisfactory
Since the single shear value governs the bolt capacity a more efficient joint would result if double-sided cover plates were used. These could be 10 mm thick, in which case four Grade 8.8 bolts would suffice (bearing now controls).

(b) *Friction-type bolted connection*
Try M20 parallel-shank HSFG bolts in clearance holes, end and edge distances are basically as for shear type (a), except that *Cl. 6.4.4* requires a minimum end distance for fully effective bearing in the plate of $3 \times 27.5 = 82.5$ mm
From *Cl. 6.4.2* for 1 interface slip capacity per bolt
$= 1.1 \times 1.0 \times 0.50 \times 144$ N $= \underline{79.2 \text{ kN}}$

From *Cl. 6.4.4* bearing capacity per bolt
= 1.5 × 20 × 20 × 460 N = 276 kN
Therefore slip capacity governs and number of bolts required
= 400/79.2 = 5.1
Use 6 bolts in 2 rows of 3.
Note A 45 mm end distance does not affect the capacity as bearing is not critical. Once again a more efficient arrangement would be to use a pair of cover plates to double the slip capacity in which case 4 bolts would be adequate.

(c) *Fillet-welded connection*
In order to accommodate the welds on the flat surface of the tie it is necessary to use a cover plate of less than 150 mm width. Since its full cross-section will be effective a 100 × 20 mm plate should be adequate (this has approximately the same area as the net section area of the plate used in cases (a) and (b)).
From *Cl. 6.7.2.2*, minimum lap length = 4 × 20 = 80 mm
From *Cl. 6.7.2.3*, if using longitudinal welds only $L \nless 100$ mm
From *Cl. 6.7.2.1*, end returns $\nless 2$ × leg length
Try 8 mm fillet welds
From *Cl. 6.8.3* throat thickness = 0.7 × 8 = 5.6 mm
Taking p_w = 220 N/mm^2
from *Table 37*
Capacity of weld per mm run = 5.6 × 220 = 1.23 kN
Therefore required length = 400/1.23 = 325 mm
Allowing for stop and start lengths according to *Cl. 6.7.2* gives a length of 325 + 2 × 8 = 341 mm
Therefore use 350 mm arranged as shown in Figure 7.17.

Exercises

1   What size fillet welds are required to attach a 150 × 12 mm flat bar hanger to the bottom flange of a 457 = 152 UB 74 so that the full tensile capacity of the hanger may be developed? Assume the use of S275 steel and electrodes for which p_w = 215 N/mm^2.
[8 mm throat size]

2   What is the capacity of an M16 Grade 4.6 bolt passing through a 12 mm plate and a 15 mm plate in (a) single shear, (b) bearing, (c) double shear assuming two 12 mm plates? Assume that the shear plane(s) pass through the threaded portion. State any conditions necessary for these strengths to be available.
[25.1 kN, 50.2 kN, 88.3 kN, end distance $\nless 32$ mm]

3   How many M24 Grade 8.8 bolts will be needed in a tension splice

comprising two 16 mm cover plates on the longer leg of a 200 × 100 × 15 mm angle in S275 steel, if the full strength of the angle is to be developed?

[6]

4   What is the shear load that can safely be carried by four M16 Grade 8.8 bolts that are already carrying 30 kN each in tension? Assume that the bolts are in single shear.

[136 kN]

5   What is the capacity of a group of four M20 parallel-shank HSFG bolts in clearance holes assuming a slip coefficient between contact surfaces of 0.50?

[327 kN]

6   Assuming that the bolt group in question 5 is subject to a shear load of 240 kN, what additional tensile load could it safely withstand?

[156 kN]

7   How many M20 Grade 4.6 bolts are needed in a tension splice on the longer leg of a 150 × 90 × 12 mm angle if the member is carrying 60% of its axial capacity? Sketch a suitable arrangement.

[10]

8   Repeat question 7 assuming the use of (a) M20 Grade 8.8 bolts, (b) M20 HSFG bolts (take $\mu = 0.45$). What length of 8 mm fillet weld would also be suitable?

[4, 6, 285 mm]

9   A 610 × 305 UB 149 is to be connected to the flange of a 350 × 368 UC 153 by a pair of web cleats using six M20 Grade 8.8 bolts in a single line on the beam web. Determine the resultant force on the most heavily loaded bolt if the vertical reaction on the beam is 540 kN. Assume 70 mm spacing between bolts and 50 mm eccentricity. Check whether this bolt is adequate.

[109 kN, unsafe in bearing in the beam web]

References

1   Cheal, B.D. (1980) *Design Guidance for Friction Grip Bolted Connections*, CIRIA Technical Note 98, London.
2   Owens, G.W. and Knowles, P.R., eds (1992) *Steel Designers Manual*, 5th ed., Blackwell Scientific Publications.
3   Owens, G.W. and Cheal, B.D. (1989) *Structural Steelwork Connections*, Butterworths, London.
4   Pratt, J.L. (1989) *Introduction to the Welding of Structural Steelwork*, SCI, London.
5   Butler, L.J. and Kulak, G.L. (1968) Strength of fillet welds as a function of direction of load, Welding Research Supplement, *Welding Journal*, 321–45.
6   Kulak, G.L., Fisher, J.W. and Struik, J.A.H. (1987) *Guide to Design Criteria for Bolted and Riveted Joints*, 2nd edn, Wiley, New York.

7 Shakir-Khalil, H. and Ho, C.H. (December 1979) Black bolts under combined tension and shear, *The Structural Engineer*, **57B**, 69–76.
8 Mann, A.P. and Morris, L.J. (1981) *Lack of Fit in Steel Structures*, CIRIA Report No. 87, London.
9 Burdekin, F.M. (1982) Tightening HSFG bolted joints, *Metal Construction*, 387–9.

Chapter 8

Joints – design

Actual connection design consists of identifying the load paths through the various parts of the connection, which must then be proportioned in such a way that an adequate margin against each possible type of failure (or limit state) is achieved. Usually this will require consideration of more than just the fastener-related modes described in Section 7.3, since features such as the ability of gusset plates to withstand the forces induced by the members they connect, the need for column webs to resist high localized compression in beam-to-column connections, etc. must also be checked.

Because of the complexity of deciding on the exact pattern of loads and stresses within a joint, for example, any attempt at rigorous analysis must include the effects of stress concentrations and localized plasticity, bolt slip, bolt preload, in-plane and bending action of the plates, local buckling, etc., it is usual to construct approximate models of joint behaviour [1–11]. Such 'models' seek to represent the main features in a manner that is sufficiently simple for rapid application in everyday design. Information of this type is not provided in BS 5950: Part 1. For most types of joint more than one acceptable model is available. This follows from the degree of simplification necessary to arrive at a workable design method being such that it can be arranged in a variety of ways, each of which fulfils the main structural requirements. Readers wishing to pursue this topic in greater depth should consult the appropriate specialist texts [1–11].

Within the UK, however, the past decade has seen a rapidly growing acceptance that many of the connections within building structures can, most efficiently, be treated by adopting a degree of standardization. Initially this was restricted to the design approach; more recently standard connections for use when nothing is to be gained by deviating from a proven arrangement have become more widely adopted. The series of BCSA/SCI 'Green Books' (1–4) present this material in the form of step by step design procedures, some technical background and tables covering standard arrangements. The 'natural' way of utilizing this in practical situations is by computerizing the process – either as spreadsheets, stand-alone programs or as part of larger CAD packages. References [1–4] cover all the primary connection types found in multi-storey and portal frames, i.e. beam-to-beam, beam-to-column, beam and column splices, column bases, eaves and apex portal connections, dealing with both 'simple construction' [1, 2] and 'continuous construction' [3] and also include the influence of composite action [4].

8.1 Beam-to-beam connections

Horizontal surfaces in structures, such as floors, are often supported on a grid of intersecting beams. Such an arrangement necessitates connections of the type illustrated in Table 8.1. Note that a prime requirement is normally that the top surface of both primary and secondary beams be at the same level. Thus several of the arrangements of Table 8.1 show notching of the end of the secondary beam – on both flanges in the extreme example of 8.1(vi). Clearly this additional operation increases the cost of fabrication; 8.1(iii), 8.1(v) and 8.1(vii) are alternatives that remove this requirement.

For all of the arrangements of Table 8.1 involving the use of bolted secondary beams it is necessary to consider a particular type of failure at the line of the holes termed block shear [12]. This is illustrated in Figure 8.1 for plain, single notch and double notch beam ends. Effectively the shaded region tears away from the rest of the beam along the line through the holes as shown. Whilst this would be covered in the case of the double notch by the ordinary shear check on a vertical line through the holes, the

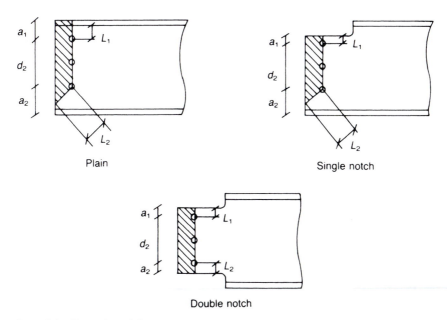

Figure 8.1 Block shear failure.

Table 8.1 Beam-to-beam connections

Joint		Design basis	Comments	Ref.
8.1 (i)		Bolts 'A' carry vertical load in shear and bearing. Bolts 'B' carry some shear plus shear due to eccentricity e of bolts from face of cleat	Use vector sum method to allow for combined loading. Alternatively use a more 'exact' ultimate load method	[1] [2] [9]
8.1 (ii)		Bolts and welds carry vertical shear only		[1] [2] [9]

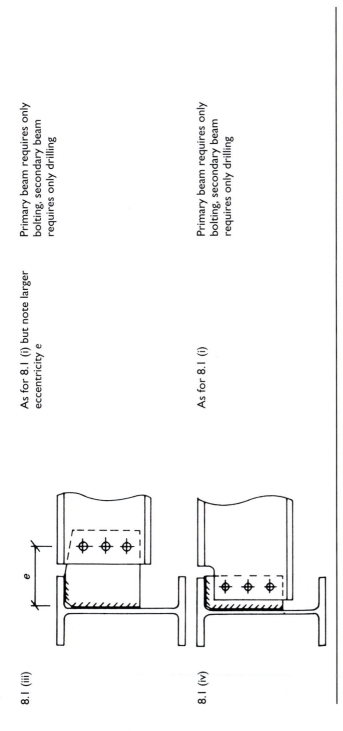

8.1 (iii)

As for 8.1 (i) but note larger eccentricity e

Primary beam requires only bolting, secondary beam requires only drilling

8.1 (iv)

As for 8.1 (i)

Primary beam requires only bolting, secondary beam requires only drilling

Table 8.1   continued

Joint		Design basis	Comments	Ref.
8.1 (v)		As for 8.1 (i)	Primary beam requires only bolting, secondary beam requires only drilling	
8.1 (vi)		As for 8.1 (i)	Primary beam requires only bolting, secondary beam requires only drilling Used when secondary beam depth exceeds that of primary, design likely to be governed by web strength of reduced cross-section	

8.1 (vii)

Tee stiffener

As for 8.1 (i)

Used where erection of (i) or
(ii) may prove difficult

—

8.1 (viii)

B

A

C

Bolts 'A' carry vertical load,
bolts 'B' transmit in shear the
couple force due to the moment

Bolts 'C' could be omitted on
the tension side (assuming
no load reversal)

[1]
[2]

region in question – using L_2 rather than a_2 – for the other two cases needs to be identified. The block shear check may therefore be expressed as

$$V = 0.6p_yA_{v,net} \tag{8.1}$$

in which $A_{v,net}$, the net area subject to block shear, is given by

$$A_{v,net} = tL_{v,eff} \tag{8.2}$$

and $L_{v,eff}$ depends on the exact pattern of holes as defined in *Cl. 6.2.4*.

Joint types (i)–(iii) of Table 8.1 are suitable when only shear is being transferred, while the heavier type (iv) is an example of a moment-resisting beam-to-beam connection. The only aspect of the design of any of these connections which has not yet been explained is the effect of the eccentricity on the bolts B of type (i). Most authorities [5, 6] recommend the use of the 'vector sum' method that is the basis for the directional method of *Cl. 6.8.7.3*; this is most easily appreciated by means of a worked example.

Example 8.1

Determine the force on the most heavily loaded bolt in the beam-to-beam connection illustrated in Figure 8.2 assuming a beam end reaction of 180 kN.

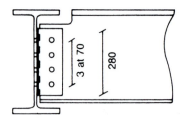

Figure 8.2

Solution
Force per bolt due to vertical shear = 180/4 = 45 kN
I for bolt group about horizontal axis of bolt group
$= 2[3.5^2 + 10.5^2] = 245$ cm^4
Z for outermost bolts = 245/10.5 = 23.33 cm^3
M due to eccentricity of line of bolts from centreline of beam
$= 180 \times 0.045 = 8.1$ kN m
Force on outermost bolts = 45 kN vertical + 8.1 × 10^2/23.33 horizontal
$\qquad\qquad\qquad\qquad\qquad\qquad = 45$ kN vertical + 34.76 kN horizontal
Resultant force = $[45^2 + 34.76^2]^{\frac{1}{2}} = \underline{56.86}$ kN

More sophisticated approaches, in which the true nonlinear load–deflection response of the bolts is used to locate the instantaneous centre of the bolt group by trial and error, are available [6, 11]. However, experimental work [13, 14] suggests that the difference in accuracy is insufficient to warrant the additional calculations. Since the presence of notches reduces the amount of lateral restraint provided [1], its effects upon the beam's overall bending strength as discussed in Chapter 5 should also be taken into account.

8.2   Beam-to-column connections – simple construction

Four examples of beam-to-column connection suitable for a frame design according to the principles of simple construction are shown in Table 8.2. Since their function is to transmit the beam reaction in shear into the column without developing significant moments, factors such as the provision of sufficient clearance between the column face and the lower flange should be properly considered. Seeking to give the end-plate protection against possible damage in transit by extending it, perhaps accompanied by welding to the beam's bottom flange, results in significant changes in the way in which the joint behaves [1]. Types (i) and (ii) are the most commonly used, the choice between them depending upon the preferred method of shop fabrication, that is whether the beam should be provided with a bolted cleat or a welded end plate. Type (iii) possesses the advantage that the seating cleat may be used for 'landing' the beam during erection (for this it must, of course, be shop welded or bolted with the site joint being made to the beam). It also possesses certain disadvantages: columns with attached cleats are less convenient for transportation, no tolerance is present to adjust for rolling margins on beam depth, etc. It is therefore not included in reference [1]. Type (iv) has gained significantly in popularity in the UK in recent years, having been used to advantage in Australia [1] and the USA [6]. It is particularly convenient for erection, permitting the beam to be swung in from one side. Because type (ii) possesses no tolerance on length, it is common practice to detail beams slightly short (1–2 mm) and to use packing to provide an exact fit.

Although each of these joints is assumed for the purposes of frame and member design to provide the equivalent of a pin support, i.e. to transfer zero moment, in reality they will each provide some (small) degree of rotational restraint and will thus attract some moment. Thus some of the bolts or part of the welds on the column flange may be expected to carry some tension. This is not normally considered in design, the justification being that the shear–tension interaction for bolts of Figures 7.10 and 7.11 show the full shear strength to be available for tensile loads up to 40% of tensile capacity. A full design treatment for end plate, web cleat and fin

Table 8.2 Beam-to-column connections suitable for 'Simple Construction'

Joint	Design basis	Comments	Ref.
8.2 (i)	Bolts 'A' carry vertical load in shear and bearing. Bolts 'B' carry some shear plus shear due to eccentricity of bolts from face of cleat	Use vector sum method to allow for combined loading or use more accurate ultimate-load methods	[1] [2] [9]
8.2 (ii)	Bolts and welds carry vertical shear only		[1] [2] [9]

8.2 (iii)

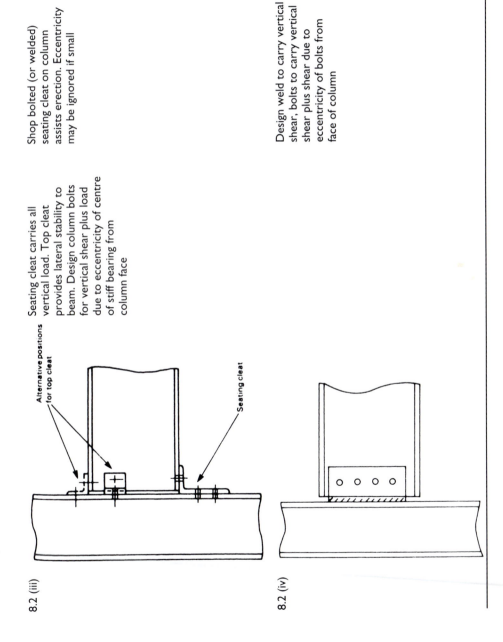

Alternative positions
for top cleat

Seating cleat

Seating cleat carries all
vertical load. Top cleat
provides lateral stability to
beam. Design column bolts
for vertical shear plus load
due to eccentricity of centre
of stiff bearing from
column face

Shop bolted (or welded)
seating cleat on column
assists erection. Eccentricity
may be ignored if small

[5]
[7]
[8]

8.2 (iv)

Design weld to carry vertical
shear, bolts to carry vertical
shear plus shear due to
eccentricity of bolts from
face of column

[1]
[2]
[9]

Heavy trusses in a Far East skyscraper.

plate arrangements is provided in reference [1]. The following two examples largely follow these procedures.

Example 8.2

Check the ability of the flush end-plate beam-to-column connection illustrated in Figure 8.3 to transfer a beam end reaction of 250 kN into the column. Both members are S275 material, the end plate is

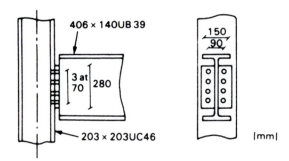

Figure 8.3

$150 \times 280 \times 10$ mm, 6 mm fillet welds are used and the bolts are M20 Grade 8.8.

Solution
The proportions of this connection follow the standard arrangement suggested in reference [1]. The following component strengths should normally be checked:

1 bolt group;
2 end plate;
3 fillet welds;
4 beam web.

(1) Bolt group
Bolt strengths according to *Tables 30 and 32*:
Shear = 375 N/mm^2; bolt bearing = 460 N/mm^2.
Bearing strength on plate according to *Table 32* = 460 N/mm^2.
From *Cl. 6.3.3.3*, end distance for full bearing strength to be developed = $2 \times 20 = 40$ mm.
Since distance provided is 35 mm, bearing capacity of last row of bolts must be reduced *pro rata*.
By inspection bolt arrangement meets requirements of *Cl. 6.2* on spacing and edge distances.
From *Cl. 6.3.2*, taking A_s as the tensile area for threads in the shear plane, capacity of bolts in single shear = $8 \times 375 \times 245 = 735$ kN
From *Cl. 6.3.3.3*, capacity of bolts in bearing on 10 mm end plate (column flange is 11.0 mm) = $6 \times 460 \times 20 \times 10 + 2 \times 460 \times \frac{1}{2} \times 35 \times 10 = \underline{713 \text{ kN}}$
Therefore bolt group capacity is controlled by bearing in end plate.

(2) End plate
From *Cl. 4.2.3*, capacity of end plate in shear = $(0.9 \times 10 \times 280) \times 0.6 \times 275 = 416$ kN
No reduction has been included for holes as it is anticipated that the thinner beam web will have a significantly smaller shear capacity.

(3) Fillet welds
From *Cl. 6.8.2*, effective length = $2[280 - (2 \times 6)] = 536$ mm.
From *Cl. 6.8.7.1*, throat thickness = $0.7 \times 6 = 4.2$ mm.
From *Table 37*, capacity per mm run = 4.2×220 N = 0.92 kN
∴ capacity of weld group = $536 \times 0.92 = \underline{493 \text{ kN}}$

(4) Beam web
From *Cl. 4.2.3* and taking $p_y = 275$ N/mm^2 from *Table 9*.
Local shear capacity of beam web = $0.6 \times 275 \times (0.9 \times 280 \times 6.3)$ N = $\underline{262 \text{ kN}}$

Summary of component capacities:

1   bolt group (bearing) 713 kN
2   end plate   416 kN
3   fillet welds   493 kN
4   beam web   262 kN.

Therefore connection capacity is limited by the ability of the beam web to transmit shear; the connection is satisfactory for the 250 kN end reaction.

Changing the number or arrangement of bolts will not improve the joint strength since it is the shear strength of the depth of beam web directly attached to the end plate that controls its capacity.

Example 8.3

Check the ability of the web cleats form of beam-to-column connection illustrated in Figure 8.4 to transfer a beam end reaction of 120 kN into the column. Both members are S275 steel, the angle cleats are $90 \times 90 \times 8$ mm and M20 Grade 8.8 bolts should be used.

Solution
The proportions of this connection follow the standard arrangements suggested in reference [1].
The following component strengths should normally be checked:

1   bolt group in beam web;
2   bolt group in column flange;
3   angle cleats in shear;
4   angle cleats in bending.

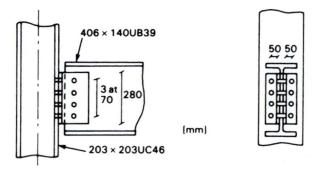

Figure 8.4

(1)   Bolt group in beam web
For 120 kN reaction, moment on these bolts = $120 \times 0.05 = 6.0$ kN m.
Using vector sum method to determine force on most heavily loaded bolt,
I for bolt group = $2(3.5^2 + 10.5^2) = 245$ cm^4,
Z for further bolts = $245/10.5 = 23.3$ cm^3.
Force on outermost bolt due to vertical shear = $120/4 = 30$ kN.
Horizontal force on outermost bolt due to moment = $6.0 \times 10^2/23.3$
= 25.8 kN.
Resultant = $(30^2 + 25.8^2)^{\frac{1}{2}} = 39.6$ kN.
From *Cl. 6.3.3.3*, end distance for full bearing strength to be developed
= $2 \times 20 = 40$ mm.
Distance beyond hole in direction of resultant bolt force = $35 \times 39.6/30.0$
= 46.2 mm.
By inspection, bolt arrangement meets requirements of *Cl. 6.2* on spacing
and edge distances.
From *Cl. 6.3.2*, and taking A_s as the tensile area since shear plane passes
through the threads, capacity per bolt in double shear = $2 \times 375 \times 245$ N
= 184 kN
Capacity per bolt in bearing in 6.3 mm beam web = $20 \times 6.3 \times 460$ N
= 58.0 kN
Since cleat thickness is 8 mm, bearing in this will be less critical.
Since capacity per bolt exceeds load on most heavily loaded bolt, group is
satisfactory.
Capacity = $(120/39.6) \times 58 = 175.8$ kN

(2)   Bolt group in column flange
From *Cl. 6.3.2*, capacity per bolt in single shear = 375×245 N = 91.9 kN
From *Cl. 6.3.3.3*, capacity per bolt in bearing in 8 mm cleat
= $20 \times 8.0 \times 460$ N = 73.6 kN
Actual capacity of last pair of bolts will be slightly less since end distance
(vertical load) of 35 mm is less than the 40 mm required; ignore this as first
approximation.
Therefore capacity of bolt group = $8 \times 73.6 = 589$ kN

(3)   Angle cleats in shear
From *Cl. 4.2.3*, shear capacity = $0.6 \times 275 \times (0.9 \times 2 \times 8 \times 280)$ N
= 665.4 kN

(4)   Angle cleats in bending
Capacity of connection = 166.0 kN.
Shear capacity of cleats = 665.4 kN.
From *Cl. 4.2.5*, since $166.0 < (0.6 \times 580.7)$, take $M_c = p_y Z$.
Gross *I* for cleats = $1.8 \times 28.0^3/12 = 3293$ cm^4
less holes $2 \times 1.8 \times 2.2 \times (3.5^2 + 10.5^2) = 970$ cm^4.

Net I for cleats = 2323 cm^4.
Z for cleats = 2323/14.0 = 165.9 cm^3.
$\therefore M_c = 275 \times 165.9$ N mm = 45.6 kN m
In terms of reaction at 50 mm eccentricity, this corresponds to a force of 45.6/0.05 = 912 kN

Summary of component capacities:

1 bolt group in beam web 175.8 kN
2 bolt group in column flange 589.0 kN
3 angle cleats in shear 665.4 kN
4 angle cleats in bending 912.0 kN.

Once again the beam web is the controlling factor; moreover, since it is bearing that controls only by using a section with a thicker web could the joint strength be made to approach more closely the strength of the components used to actually make the connection, i.e. the bolts and the cleats.

8.3 Beam-to-column connections – continuous construction

This form of connection may be made in a great variety of ways, six of which are illustrated in Table 8.3. Before considering these in detail it will be useful to establish certain points relating to the design of moment-resisting connections in general.

1 The beam end moments will also contribute to the shear force at the joint.
2 Axial tension or compression may be present in the beam; its effect on connection design should be approached with caution since such forces may well be present only under certain conditions of loading.
3 Tension in the beam will, as a result of rotation of the joint, produce additional moment. If this effect is significant, placing of the bolts symmetrically with respect to the resultant line of action of the applied forces enables them to be designed for tensile forces only, thereby assisting in keeping connection size reasonable.
4 Compression in the beam, since it has the opposite effect, can lead to lighter connections.
5 The compression zone of the column web should be checked for possible failure in local bearing and buckling (see Chapter 5); some stiffening may be necessary [4–8, 15–17].

Table 8.3 illustrates six examples of moment-resisting beam-to-column joints suitable for use in continuous construction. The most popular of

Table 8.3 Beam-to-column connections suitable for 'continuous construction'

Joint	Design basis	Comments	Ref.
8.3 (i)	Bolt tension calculated by assuming beam to rotate about its compression flange, top row at least assumed at yield. Divide shear between all bolts (or assume taken by bottom row only). End-plate design assumes double-curvature bending	Prying action present but not normally considered (some allowance in bolt strengths. Column web may need stiffening	[3] [6] [9] [15] [18]
8.3 (ii)	Generally as for 8.3 (i). Haunch flange carries compression force as a strut ($l \approx 0.7L$)	Haunches often cut from same size UB as main member	[3] [15] [18]

Table 8.3 continued

Joint		Design basis	Comments	Ref.
8.3 (iii)		Webs bolts take the shear, bolts 'A' (acting in shear) resist the moment	Cover plate may be supplied loose for site bolting to a welded cap plate	[5] [17]
8.3 (iv)		Bolts carry both shear and tension	Only four bolts may be used. For heavy shears use a shear pad welded to the column flange toes	[17]

8.3 (v)

Bottom cleat takes whole shear, top cleat provides the tensile resistance to develop the moment capacity

Unsuitable for large moments

8.3 (vi)

Web cleats take all the shear, moment is resisted by the couple force developed in the flange connections

Tee-stubs usually cut from UBs

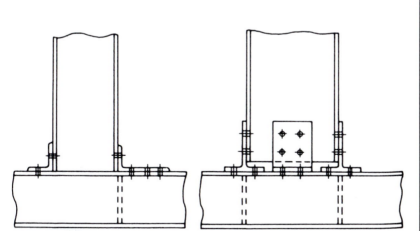

these is type (i), the extended end plate. Variants of this are possible in which the end plate is made almost flush with the bottom of the beam (assuming downward loading on the beam) or even when it is effectively contained within the beam depth, although evidence [15] suggests that for the latter case very thick plates are necessary to resist the induced moments (equal to the product of the beam flange force and its distance from the nearest row of bolts). A detailed treatment of end-plate connection design is provided in reference [3].

End-plate connections have been the subject of considerable study in recent years, with the result that a generally accepted design procedure that closely approximates the mechanics of load transfer through the various components in the actual joint is widely used in the UK. Both the physical background to this and its application in a rigorous fashion – intended for computer implementation – and in a simpler form – using design aides – are presented in [3]. The origins for this approach are research conducted in the Netherlands. Although it appears in Eurocode 3, the treatment given in [3] includes many refinements to make it compatible with British practice and to permit detailed strength checks to be undertaken using the provisions of BS 5950: Part 1.

The method requires the making of the fifteen checks listed in Table 8.4 to cover the fifteen potential modes of failure illustrated in Figure 8.5. Most are straightforward but the determination of the tension forces transmitted by each bolt row requires care since at each level it is linked to the ability of the adjacent plating on each side of the connection, i.e. column

Table 8.4 Design checks for end-plate connection (after reference 3)

Zone	Ref.	Checklist item	See procedure
Tension	a	Bolt tension	Step 1A
	b	End-plate bending	Step 1A
	c	Column flange bending	Step 1A
	d	Beam web tension	Step 1B
	e	Column web tension	Step 1B
	f	Flange to end-plate weld	Step 7
	g	Web to end-plate weld	Step 7
Horizontal shear	h	Column web panel shear	Step 3
Compression	j	Beam flange compression	Step 2
	k	Beam flange weld	Step 7
	l	Column web crushing	Step 2
	m	Column web buckling	Step 2
Vertical shear	n	Web to end-plate weld	Step 7
	p	Bolt shear	Step 5
	q	Bolt bearing (plate or flange)	Step 5

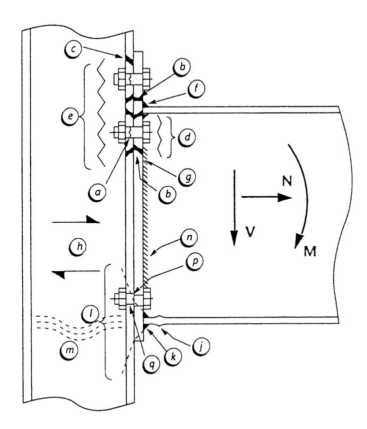

Figure 8.5   Potential failure modes.

flange and beam end plate to transmit the bolt force. Since it is normal practice to reserve the bottom row of bolts to transmit vertical shear only – even if they could generate a worthwhile tensile force, the lever arm from the assumed point of compressive force transfer at the beam bottom flange means that they could only contribute a very small percentage to the overall moment capacity of the connection – consideration needs to be given to determining a suitable set of forces for each bolt row adjacent to the beam's tension flange. Following normal structural principles bolts in stiffer parts of the system, i.e. those immediately adjacent to the beam flange, are assumed to contribute most.

When determining bolt row resistances the approach employs the concept of an equivalent tee-stub to deal with the interaction between the bolts and the adjacent plating. The concept of a tee-stub is illustrated in Figure 8.6. The appropriate representation for any particular bolt row will depend upon its precise location, in particular the extent to which it benefits from support provided by a nearby flange, stiffener etc, as explained in

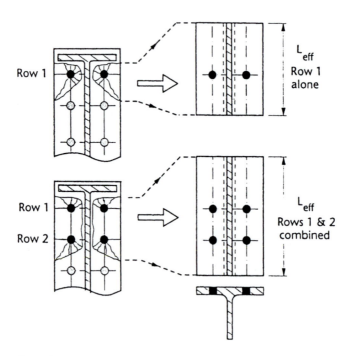

Figure 8.6   Tee-stub.

[3]. Three possible modes of failure have been identified for tee-stubs and these are shown in Figure 8.7. Mode 1 is associated with large plate deformations and a ductile failure. Mode 2 involves interaction between bolt stretch (and bending) and plate deformation but is still reasonably ductile. Although Mode 3 utilizes the full bolt tensile resistance, it involves an undesirable brittle failure. The design intent is, therefore, to produce a connection governed by either Mode 1 or Mode 2. An important feature of this approach, is that the resulting equations for strength checking

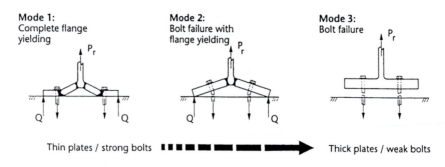

Figure 8.7   Potential failure modes.

recognize the extent to which prying is present, both for each mode and within a mode for the particular combination of plate thicknesses, geometrical arrangement etc. Thus the full tensile bolt resistances of *Table 34* should be used.

In its full form the method is only really suitable for implementation using computer software; several standard packages exist. A simpler, approximate version is also presented in [3]. This omits much of the process whereby bolt row forces are adjusted, so as to produce the most effective arrangement in terms of maximizing moment capacity. It still requires a significant amount of calculation – largely because each of the checks listed in Table 8.4 must be explicitly conducted.

Thus readers wishing to pursue this topic are strongly advised to study the presentation in [3]. A step by step treatment and explanation of both the full and abridged method, illustrated by worked examples, is provided.

In situations where the moment at the joint exceeds the capacity of the beam section, a haunched connection of the type shown in Table 8.3 as 8.3 (ii) may be used, a common example being the eaves of a portal frame (see Chapter 10). Haunches may be made either from split UB sections or from plate.

At a column cap the type of joint shown as 8.3 (iii) is suitable. An alternative, all-welded arrangement would be to run the beam through the connection and to use vertical stiffeners to extend the column flanges. A design model based on North American practice is provided in reference [17].

Although types (i) and (ii) are also suitable for beams framing into the column web this may present difficulties if moment connections are required on both axes. Type 8.3 (iv) represents one means of making such a joint by employing tee-stiffeners to effectively move the connection to the column face. Such stiffeners will, of course, also act to stiffen the column web against major axis bending.

The top and bottom cleat arrangement used previously as a simple connection can be used to transmit moments providing the bottom cleat is made much more substantial, which in turn will probably require stiffening of the adjacent column web. A cleated connection capable of transmitting large moments is shown in 8.3 (vi). This uses tee-stubs cut from UBs as the flange connections. Since these are symmetrically loaded they deform less than the eccentrically loaded angles of type (v); they also permit the use of more bolts.

Not shown in Table 8.3 are any all-welded joints. Structurally, these represent the simplest form of moment-resisting beam-to-column connection. However, this must be balanced, not only against the need to employ site welding, but also against the generally rather higher degree of precision necessary in fabrication and fit-up. Nonetheless, such connections are sometimes used in the UK; they are much more common in regions such as North America and Japan, where greater use is made of continuous construction. For a discussion of their design, which requires that careful consideration be given to factors such as ductility and the provision of adequate stiffening, the reader is referred to references [5, 8, 16, 18]. A detailed procedure based on the use of BS 5950: Part 1 component strength rules is given in reference [3].

8.4   Column splices

Joints between successive parts of columns are necessary if individual column lengths are to be kept within manageable proportions. Although such splices provide an opportunity for changing column cross-section, only limited use is normally made of this as it is often more economic and practically more convenient to rationalize on a small number of section sizes throughout the project. Apart from special situations, for example where heavy additional loads must be carried over only a portion of the column's height, as occurs with crane columns, it is usual practice to retain common outside dimensions over the full column height.

For cases of predominantly axial loading either of the two arrangements of Figure 8.8 may be used. Both are designed to transmit principally compressive load but do so in different ways.

In the direct bearing arrangement of Figure 8.8 (a) the ends of both sections are assumed to make sufficiently good contact that the whole of the load is transferred through the contact area. The splice plates are there for location, as a safeguard against any accidental lateral forces and possibly to withstand any direct tension if the splice has to be capable of resisting limited tensile force (as is often required nowadays because of the possibility of uplift loading from internal explosions in buildings). The ends of

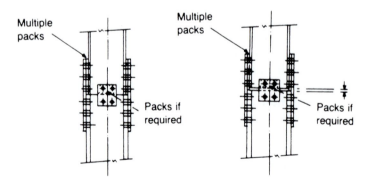

Figure 8.8   Alternate forms of column splice: (a) Column splice, ends prepared for direct bearing. (b) Column splice ends not prepared for direct bearing.

both columns may require machining, i.e. milling, although as equipment improves it is now common practice for the cuts produced by a good quality, well-maintained saw to be quite acceptable. BS 5950: Part 2 gives guidance on the level of tolerance required.

As an alternative a gap may be left between the member ends and the whole of the load transferred by means of the splice plates. Clearly these will now need to be more substantial with considerably more bolts being used. The quality of fabrication is, of course, less important as the ends will not be in contact.

For either case the splice plates may be located on the inside of the column flanges so as to reduce the plan area occupied by the column.

If load reversal is possible, end-plate connections provide a convenient solution as shown in Figure 8.9 (a). High-tensile or possibly HSFG bolts are usual and care is necessary in selecting material free from laminations for the end plate due to the tensile loading involved.

Connections between columns of very different size may be arranged as shown in Figure 8.9 (b); both faces of the division plate should be machined and its thickness will normally need to be at least 20 mm. The web stiffener, which assists in diffusing load into the lower column, should be of similar proportions to the upper column flange.

Column splices in a multistorey frame are normally required at something between every second and every fourth floor. With typical storey heights of 3.5–4 m this gives manageable lengths of up to about 16 m, compared with readily obtainable lengths of the standard rolled sections of at least 20 m. It is normal practice to position splices just above floor level so that the effects of flexing of the column may be neglected. For splices located in regions subject to column flexure *Cl. C.3* provides the means to calculate the necessary additional bending effects.

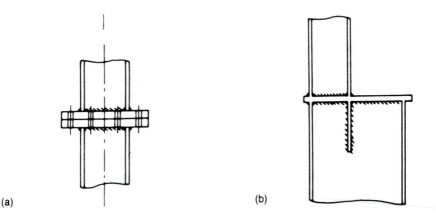

(a)   (b)

Figure 8.9 Column splices.

Example 8.4

Check the ability of the column splice illustrated in Figure 8.10 to transfer a combination of forces corresponding to a direct compression of 500 kN, a moment of 120 kN m and a horizontal shear of 30 kN. Assume that the splice is designed for direct bearing and that M20 bolts are to be used. All material is S275.

Solution
The following components should normally be checked:

1   cover plate;
2   bolt group.

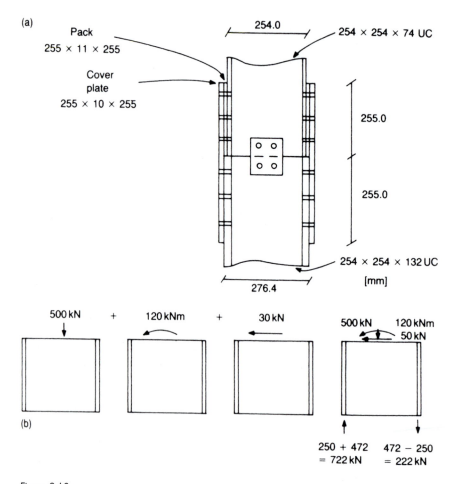

Figure 8.10

Item (2) is required only if tension can be developed; the first check should therefore be on the design forces for each side of the splice as shown in Figure 8.10 (b).

(1) Cover plate
From *Cl. 4.6.1* tensile capacity = Ap_y
in which A is given by *Cl. 3.4.3* as the lesser of K_eA_n or A_g
$K_eA_n = 1.2(255 - 2 \times 22)10 = 2532$ mm^2
$A_g = 255 \times 10 = 2550$ mm^2
Capacity of cover plates = $2532 \times 275 = \underline{696\ kN}$

(2) Bolt group
For one M20 bolt in single shear, using *Table 30*, capacity = 375×245 = 91.9 kN.
Assuming no reduction for insufficient end distance capacity of group = $6 \times 91.9 = \underline{551\ kN}$
For one M20 bolt in bearing in 10 mm plate, using *Table 32*, capacity = $460 \times 20 \times 10 = 92$ kN.
Capacity of group = $6 \times 92 = \underline{552\ kN}$

Summary of component capacities

1	cover plate	696 kN;
2	bolt group (shear)	551 kN.

Therefore connection is quite safe for combination of axial load and moment since tensile load is 222 kN. The small horizontal shear may readily be accommodated by friction at the interface.

8.5   Beam splices

Long-span beams may require site connections between successive lengths. Figure 8.11 illustrates two basic forms of beam splice, both of which can have several variants. For the end-plate arrangement, design is similar to that discussed previously for the beam-to-column end plate, i.e. shear is assumed to be shared equally between all bolts with the moment being resisted by a group of tension bolts. The flange cover plates in Figure 8.11 (b) should be capable of transmitting the whole of the moment with the web bolts taking shear plus the secondary moment due to their eccentricity. For large beams, flange plates may be placed on both faces of the beam flanges; HSFG bolts will often be required if the number of bolts used is to remain reasonable. As an alternative, welded splices may be used to provide a particularly clean appearance. Reference [2] provides detailed discussion of various design approaches together with example calculations for a number of different types of beam splice.

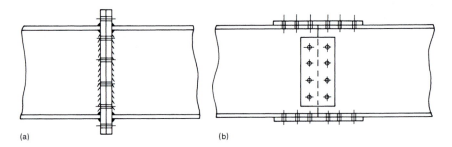

Figure 8.11   Beam splices: (a) end plates, (b) cover plate.

In the example that follows the web splice bolts have been designed for the vertical shear plus a moment assumed to be given by the product of this force and the distance from the bolt row to the centre-line of the splice.

This assumption is by no means universally agreed upon. For example both the 5th edition of the *Steel Designers' Manual* [8] and reference [5] assume the line of action of the bolt group to be the centroid of the opposite bolt group, i.e. use twice the distance. It is thus of interest to note that a recent more rigorous theoretical study [20], supported by a limited number of large scale tests, has confirmed the use of the present approach as the most suitable.

Example 8.5

Check whether the beam splice illustrated in Figure 8.12 is capable of transmitting a moment of 159 kN m together with a shear of 250 kN m. Flange cover plates are 15 mm and web cover plates are 8 mm. All bolts are M20 general grade HSFG and all material is S275.

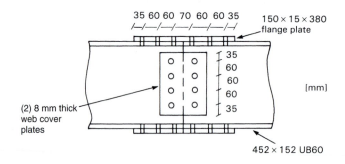

35 60 60 70 60 60 35

150 × 15 × 380 flange plate

[mm]

(2) 8 mm thick web cover plates

452 × 152 UB60

Figure 8.12

Solution
The proportions of this connection are in accordance with those suggested
by reference [1]. Design is based on the assumption that moment is transmitted entirely by the flange plates with the web plates carrying the whole
of the shear. Items to be checked:

1 web splice bolts ⎫ Compare with resultant
2 web cover plates in shear ⎬ force due to shear + moment
3 web cover plates in bending ⎭ due to eccentricity
4 flange splice bolts ⎫ Compare with
5 flange cover plates in compression ⎬ flange force due
6 flange cover plates in tension. ⎭ to moment

(1) Web splice bolts
Vertical shear on bolt group = 250 kN
Moment due to eccentricity = $250 \times 0.035 = 8.75$ kN m
Force on outermost bolts:

Due to shear $= 250/4$ $= 62.5$ kN

Due to moment $= \dfrac{8.75 \times 13.5 \times 10^2}{2(4.5^2 + 13.5^2)}$ N $= 29.1$ kN

Resultant force $= (62.5^2 + 29.1^2)^{\frac{1}{2}} = 68.9$ kN
From *Cl. 6.4.2*, assuming $\mu = 0.50$ and noting that two interfaces are
present,
slip resistance of one bolt $= 1.1 \times 1.0 \times 0.50 \times 144 \times 2 = 158.4$ kN.
From *Cl. 6.4.2.2*, noting that $e = 35 \times (69.9/29.1) > 3d$,
bearing resistance in 8 mm beam web $= 20 \times 8 \times 825$ N $= 132$ kN.
For the 16 mm of cover plate, $e = 35 \times (68.9/62.5) < 3d$, but capacity is still
>132 kN.

Therefore capacity of bolt group in shear $= \dfrac{250 \times 132}{68.9} = \underline{480.0 \text{ kN}}$

Since this exceeds 250 kN applied this item is satisfactory.

(2) Web cover plates in shear
From *Cl. 4.2.3* and *Table 9*, shear capacity of cover plates
$= 0.6 \times 275 \times (0.9 \times 8 \times 340) \times 2$ N
$= \underline{807.9 \text{ kN}}$ (satisfactory).

(3) Web cover plates in bending
Since capacity of web splice bolts (480.0 kN) $< 0.6 \times$ shear capacity of web
cover plates (485 kN)
from *Cl. 4.2.5*, take $M_c = p_y S \not> 1.2 p_y Z$
For a rectangular plate, $S = bd^2/4$ and $Z = bd^2/6$

∴ use M_c = 1.2 × 275[16 × 340³/12 − 16 × 22 × 2(45² × 135²)]/(170 × 35)
        = 2115 kN (satisfactory).

Note   In determining Z, allowance has been made for the presence of holes.

(4)   Flange splice bolts
Force taken by bolt group on either side of splice = 150/0.455 = 330 kN.
From *Cl. 6.4.2*, taking μ = 0.50 for one interface,
P_s = 1.1 × 1.0 × 0.50 × 144 = 79.2 kN.
From *Cl. 6.4.3*, for bearing in 13.3 mm flange plate, noting that e = 35 mm,
P_{bg} = 1/3 × 35 × 13.3 × 825 N = 128 kN
∴ capacity of bolt groups = 6 × 79.2
                = 475.2 kN (satisfactory).

(5)   Flange plate in compression
For top plate from *Cl. 4.7.4*, $P_c = A_g p_c$.
Gross area A_g = 22.5 cm².
Noting that close spacing of bolts will give a low slenderness so that $p_c \approx p_y$, capacity of cover plates in compression = 2250 × 275 = 619 kN (satisfactory).

(6)   Flange plate in tension
From *Cl. 4.6.1*, P_t = 525.0 kN as before (satisfactory).

Summary of component capacities:

1	Web splice bolts	480 kN
2	Web cover plates in shear	808 kN
3	Web cover plates in bending	2115 kN
4	Flange splice bolts	475 kN
5	Flange cover plates in compression	619 kN
6	Flange cover plates in tension	619 kN.

Items (1)–(3) exceed the load produced by the shear while items (4)–(6) are capable of resisting the 330 kN flange force produced by the moment.

8.6   Column bases

Transfer of column loads into masonry or concrete foundations usually requires the insertion of a steel plate between the two components if over-stressing of the weaker foundation material is to be avoided. Adjustment of level is facilitated by the insertion of cement grout between the under-side of the baseplate and the top of the concrete. This grout layer is likely to be of a significantly lower strength – say one quarter to one half – that of the concrete foundation. For columns carrying only axial load, direct bearing between the column end and the top of the plate may be used to transmit compression. The welds shown in Figure 8.13 are then used only for location or perhaps to transfer any small shears or tensions that might develop under particular load combinations. This arrangement may require the contact surfaces to be machined. As an alternative, usually when only small loads are involved, machining may be omitted, with the whole of the load being transferred by the welds.

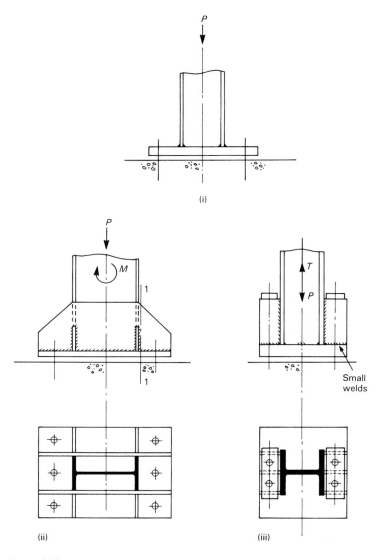

Figure 8.13   Column bases: (a) slab base, (b) haunched base, (c) bolt boxes.

Clause 4.13 of BS 5950: Part 1 permits baseplates to be proportioned using any rational method; it also contains a method based on the use of an effective area of baseplate.

The concept, which is illustrated in Figure 8.14, gives the minimum required plate thickness as:

$$t_p = c[3w/p_{yp}]^{\frac{1}{2}} \qquad\qquad (8.3)$$

in which c = maximum perpendicular distance from the edge of the effective portion of the baseplate to the face of the column cross-section.

p_{yp} = design strength of the baseplate
W = pressure under the baseplate

The approach is based on an effective bearing area over which the pressure w is assumed to be uniform and limited to two thirds of the design value of the concrete cylinder strength of the foundation.

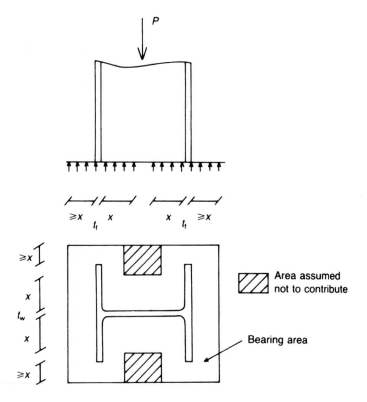

Figure 8.14 Definition of effective bearing area.

Baseplates for columns designed to transmit significant moments into the foundations may need haunching as shown in Figure 8.13 (b). *Clauses 4.13.2.3* and *4.13.2.4* explain how the effective area approach may be adapted to cover these items.

Cases where baseplates are required to transmit large tensile forces entail the use of very thick plates to resist the moments produced by the holding-down bolts. In combination with heavy welding this can lead to lamellar tearing [6] in the baseplate. One way of avoiding it is to modify the method of load transfer by using bolt boxes as shown in Figure 8.13 (c). Most of the load is now carried by the fillet welds between the boxes and the column flanges.

Information on the selection and design of a suitable holding-down system may be found in the publication produced jointly by BCSA, the Steel Construction Institute and the Concrete Society [21]. Alternatively manufacturers of proprietary systems normally produce their own technical literature giving design guidance. Figure 8.15 illustrates two of the more basic anchoring devices. Readers wishing to learn something of the various structural interactions that govern the design of holding-down systems should consult the test report by Ueda, Kitipornchai and Ling [22].

8.7 Truss connections

The joints required in trusses and lattice girders are of a somewhat different kind [10, 11] from those considered so far; Sections 8.1–6 have dealt with the various types of connection required in an essentially rectangular beam and column framework. Triangulated framing differs in requiring other than right-angled connections between members subject principally to axial forces.

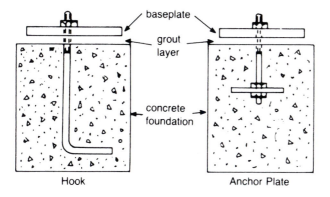

Figure 8.15 Typical holding-down systems.

One common requirement is for splices as illustrated in Figure 8.16, where the principles of designing a splice in a tension or compression boom that is too long for economic transportation as a single length are basically the same as those used to design column splices. Depending on both the exact location of the splice and the presence of loads between main truss joints, some bending may also be present.

Another, rather more fundamental, difference occurs when closed sections are used. In addition to being structurally efficient when carrying largely axial loads, structural hollow sections (SHS) offer clean lines and are therefore visually attractive. If the joints are to preserve this clean appearance they should not be visually intrusive; this virtually dictates the use of welded connections. Moreover, problems of access make the devising of acceptable mechanical connections difficult with the result that tubular trusses tend to be of all-welded construction. Figure 8.17 illustrates a number of basic joint types.

8.7.1   Open section trusses

In situations where a number of differently oriented members meet at a joint, the transfer of forces between them may be achieved conveniently by means of a piece of plate termed a gusset. Figure 8.18 illustrates an example of the type often seen in fairly light roof trusses.

Ideally the centroids of all members, and thus the lines of action of the direct tensile or compressive forces in them, should intersect at a point. However, the practicality of actually making the connection may not permit this, in which case the effects of eccentricity of loading should be allowed for in its design. Although this is often neglected for light to medium trusses, proper allowance should be made when designing the connections between heavily loaded members, for example joints in large trusses fabricated from universal column sections of the sort used in power stations and bridges.

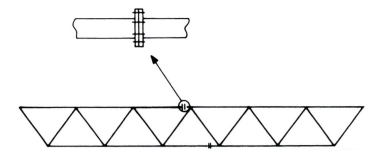

Figure 8.16   Tubular truss showing bolted flange plate type of splice in chord members.

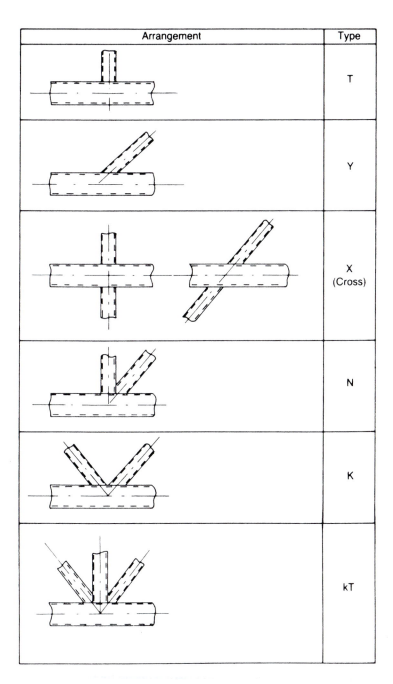

Arrangement	Type
	T
	Y
	X (Cross)
	N
	K
	kT

Figure 8.17 Basic types of tubular joints.

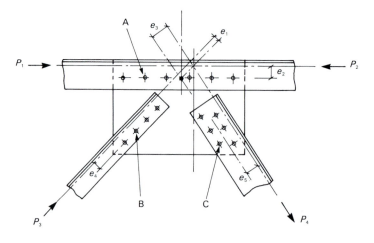

Bolt group	Direct load	Moment
A	$P_2 - P_1$	$(P_3 e_1) + (P_4 e_3)$ $- (P_2 - P_1)e_2$
B	P_3	$P_3 e_4$
C	P_4	$P_4 e_5$

Figure 8.18   Gusseted connection.

The gusset plate itself will normally be subjected to bending, shear and axial force components induced as a result of shear transfer through the fasteners from the members. Thus design of the actual gusset consists essentially of checking a number of critical sections, usually at bolt positions as indicated in Figure 8.18. Even in cases where the load application is not symmetrical with respect to the gusset plate, as with a gusset connecting a series of single angles, the joint geometry is often such that out-of-plane bending is insignificant; it is therefore usual to neglect this in design.

One problem concerns the amount of gusset that may reasonably be assumed to carry the load transferred by any particular member. A simple way of dealing with this is to use the concept of effective width illustrated in Figure 8.19 [9]. Suitable reductions should be employed so as to avoid overlapping of the effective widths in regions where members are closely spaced.

Gusset plates may be omitted, thereby reducing the labour content of fabrication, providing the chord members have sufficient space to permit fastening (welds or bolts) to take place directly on the member.

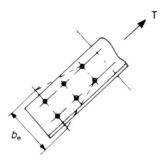

Figure 8.19   Definition of effective width b_e of gusset plate.

8.7.2   Tubular trusses

Each of the basic arrangements of Figure 8.17 may be produced by welding around the end of the web member(s). In the case of rectangular hollow sections (RHS) ends may be cut straight (but obliquely) unless the overlap arrangement of Figure 8.20 is used. Circular hollow sections (CHS) on the other hand will require complex profiling of the end of the incoming member whatever form of intersection is used.

Various possible modes of behaviour and therefore a series of potential failure modes are possible with SHS joints depending upon:

1   layout – T, Y, K etc.;
2   member type (RHS or CHS), size and proportions, especially the wall thickness;
3   applied loading;
4   detailing – gap, overlap etc.

The provision of comprehensive, detailed guidance is therefore beyond the scope of BS 5950: Part 1. Design rules for a wide variety of cases have been published by CIDECT [23], some of these are contained in references [9] and [11]. Much of the background to these rules, together with

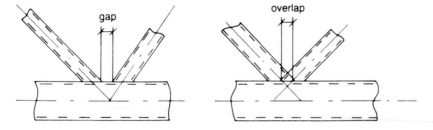

Figure 8.20   Definition of gap and overlap cases.

explanations of the physical phenomena involved, is provided in the text by Wardenier [24].

8.8 Bracing connections

Bracing is frequently used (see Chapter 10) in beam and column type structures as a means of providing enhanced lateral stiffness. Thus a limited number of diagonal members is added to the basic rectangular layout in the vertical and/or horizontal planes so as to provide some triangulated regions. This requires some modification to the beam-to-column connections as illustrated in Figure 8.21. Providing the particular arrangement used maintains concentric member centre-lines, then the additional forces for which the individual connection components must be designed are readily determined.

Constructing the free body diagram of Figure 8.21 (b) and resolving the tensile force F in the upper diagonal into horizontal and vertical components permits forces at the various member interfaces to be determined as:

welds between gusset and beam	horizontal shear	$F \cos \alpha$
welds between gusset and column	vertical shear	$F \sin \alpha$
bolts between end plate and column	vertical shear plus the end reaction due to beam loads.	$F \sin \alpha$

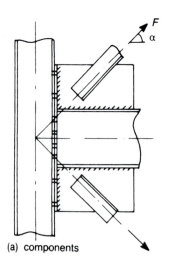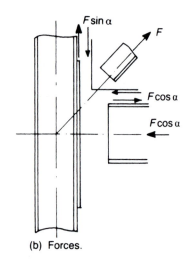

(a) components

(b) Forces.

Figure 8.21 Bracing connection, (a) components, (b) forces.

This neglects the effect of the eccentricity of half the column width in producing a moment on the column flange bolts equal to the product of the total vertical shear times the eccentricity. Shifting the point of intersection of the member axes to the column face would eliminate this – but would, of course, cause the same effect to be transferred to the column.

A particularly good explanation of various possible approaches to the design of bracing connections may be found in the paper based on American practice by Thornton [25].

8.9   Structural integrity of connections

Clause 2.4.5 of BS 5950: Part 1 places certain requirements on steel frame structures in terms of their ability not to suffer disproportionate collapse in the event of localized damage being caused by abnormal loading. The subject is considered fully in Chapter 10 in the light of a recent interpretation of the code rules. Integrity considerations have a direct influence on connection design in that the tying action of beams requires the connections to possess adequate direct tensile capacity.

Recent experimental work [26] has shown that the requirements for all buildings of resisting a factored tensile load of 75 kN (40 kN at roof level) may readily be met by both end plates and web cleats of 8 mm thickness fastened to the column flange by top M20 Grade 8.8 bolts. Recognizing that this is an ultimate condition that need not be considered in combination with other load cases, a design approach that utilizes both the ultimate material strength and the more favourable deformed geometry at failure is available [1] to cope with those situations in which larger forces must be designed for.

Exercises

1   Check whether an 8 mm thick end plate in S275 steel of less than the full beam depth used in conjunction with ten M20 Grade 4.6 bolts, would be a suitable connection between a 457 × 191 UB 74 beam and a 254 × 254 UC 89 column if the beam end reaction is 300 kN.

[Suitable]

2   A pair of 90 × 90 × 10 mm angles are to be used as web cleats to form a connection between a 610 × 229 UB 113 and a 254 × 254 UC 132 using six M20 Grade 8.8 bolts in a single line in the beam web. Is this a safe arrangement for a beam reaction of 400 kN?

[Yes]

3   Check the suitability of a 16 mm thick extended end plate welded to a 406 × 178 UB 54 with 6 mm fillet welds and fastened to the flange of a 254 × 254 UC 74 with six M20 Grade 8.8 bolts to carry a shear of 90 kN and a moment of 80 kN m.

[Suitable]

4   Check whether a flush end plate welded to both flanges of the beam but with all the bolts contained within the beam depth can be used as an alternative solution for question 13.

[Yes]

5   Design a baseplate for a 305 × 305 UC 198 assuming that the column has been designed as 'pin-ended'.

[700 × 700 × 50 mm with four M24 Grade 4.6 nominal bolts and 8 mm nominal fillet welds would be suitable]

6   A 457 × 152 UB 60 beam requires a splice at a point where the shear is 260 kN and the bending moment is 150 kN m. Assuming the use of 8 mm web cover plates and 15 mm flange cover plates with M20 general grade HSFG bolts, design a suitable joint.

[A possible arrangement would be 350 × 150 × 15 mm flange plates, 340 × 140 × 8 mm web plates, 8 web bolts and a total of 24 flange bolts]

References

1 BCSA/SCI (1993) *Joints in Simple Construction, Vol. 1: Design Methods*, 2nd edn, SCI P-205.
2 BCSA/SCI (1992) *Joints in Simple Construction, Vol. 2: Practical Applications*, SCI P-206.
3 BCSA/SCI (1995) *Joints in Steel Construction: Moment Connections*, SCI P-207.
4 BCSA/SCI (1998) *Joints in Steel Construction: Composite Connections*, SCI P-213.
5 Owens, G.W. and Cheal, B.D. (1989), *Structural Steelwork Connections*, Butterworths.
6 Kulak, G.L., Fisher, J.W. and Struik, J.H.A. (1989) *Guide to Design Criteria for Bolted and Riveted Joints*, 2nd edn, Wiley.
7 Thornton, W.A. and Kane, T. (1997) 'Connections', chapter 7 of *Steel Design Handbook – LRFD Method*, McGraw-Hill.
8 Owens, G.W. and Knowles, P.R. eds (1992) *Steel Designers' Manual*, 5th edn, Blackwell Scientific Publications.
9 Hogan, T.J. and Thomas, I.R. (1994) *Design of Structural Connections*, 4th edn, AISC.
10 Syan, A.A. and Chapman, B.G. (1996) *Design of Structural Steel Hollow Section Connections*, AISC.
11 Packer, J.A. and Henderson, J.E. (1999) *Hollow Structural Section Connections and Trusses*, 2nd edn, CISC.
12 Birkemoe, P. and Gilmor M.I. (1978) 'Behaviour of Bearing Critical Double Angle Beam Connections', *Engineering Journal AISC*, **15**(4), 109–15.
13 Crawford, S.F. and Kulak, G.L. (1971) 'Eccentrically Loaded Bolted Connections', *Journal of the Structural Division*, ASCE, 97 (ST3), 765–83.
14 Bahia, C.S. and Martin, L.H. (1980) 'Bolt Groups Subject to Torsion and Shear', Proceedings Institution of Civil Engineers, **69**(2), 473–90.
15 Horne, M.R. and Morris, L.J. (1981) *Plastic Design of Low-rise Frames*, Granada, London.
16 Chen, W.F. and Lui, E.M. (1988) 'Flange Moment Connections', in W.F. Chen

(ed.) *Steel Beam-to-Column Building Connections*, Elsevier Applied Science, pp. 39–88.

17 American Society of Civil Engineers (1971) *Plastic Design in Steel*, Manual 41, 2nd edn, ASCE.

18 Morris, L.J. (1988) 'Design Rules for Connections in the United Kingdom', in W.F. Chen (ed.) *Steel Beam-to-Column Building Connections*, Elsevier Applied Science, pp. 375–415.

19 Chen, W.F. and Lui, E.M. (1988) 'Static Web Moment Connections' in W.F. Chen (ed.) *Steel Beam-to-Column Building Connections*, Elsevier Applied Science, pp. 89–132.

20 Kulak, G.L. and Green, D.L. (1990) *Design of Connectors in Web-flange Beam or Girder Splices*, *AISC Engineering Journal* (second quarter), 41–8.

21 SCI, BCSA, Concrete Society (1980) 'Holding Down Systems for Steel Stanchions'.

22 Ueda, T., Kitipornchai, S. and Ling, K. (1988) 'An Experimental Investigation of Anchor Bolts Under Shear', Department Civil Engineering, University Queensland, Research Report CE93, October.

23 Giddings, T.W. and Wardenier, J. (eds) (1986) Monograph No 6, 'The Strength and Behaviour of Statically Loaded Welded Connections in Structural Hollow Sections', CIDECT.

24 Wardenier, J. (1982) *Hollow Section Joints*, Delft University Press.

25 Thornton, W.A. (1991) 'On the Analysis and Design of Bracing Connections', AISC National Steel Construction Conference, Washington, pp. 26–1 to 26–33.

26 Owens, G.W. and Moore, D.B. (1990) *Outstanding Problems in Simple Building Connections*, Welded Structures '90, Welding Institute.

Chapter 9

Composite construction

Apart from certain, rather specialized types of structure, e.g. transmission towers, cranes, plant supports, etc., steelwork does not normally exist in isolation – despite a far too frequent but altogether misguided tendency for it to be designed as if it had no real interaction with anything else. However, one area in which the potential benefits of properly considering the combination of the steel frame with other structural elements is appreciated is in so-called composite construction. In this case the combination is between steel and reinforced concrete, although to some extent the concept is merely an extension of the more basic idea of reinforced concrete. The principal difference is that steel sections capable of carrying significant load in their own right are used in composite construction; in conventional reinforced concrete the reinforcement is, of course, not really capable of functioning on its own as a structural element.

The essential features of composite construction may best be appreciated by considering its most widely used application: the composite beam. Figure 9.1 illustrates the concept of a beam consisting of two constituent parts acting either separately or compositely. For the present the particular materials or proportions do not matter; the key aspect is the difference in the mechanics of load resistance.

For the non-composite arrangement the load will be shared between the two parts with each deforming in bending and generating separately the typical linear variation of strain over its own depth. Now consider the same arrangement but with continuity preserved along the horizontal interface so that both parts respond as a unit. Bending strains will now vary linearly over the whole depth, with the neutral axis for the combined

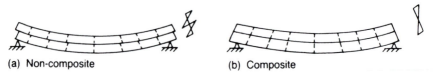

(a) Non-composite   (b) Composite

Figure 9.1 Mechanics of composite action.

section corresponding to the locus of zero strains. Moreover, since no horizontal slip will occur at the interface, vertical lines drawn on the depth of the section before loading will remain as single lines as shown. Clearly the composite arrangement may be expected to be more efficient structurally, developing smaller deflections and smaller strains than its non-composite equivalents. If both parts were of the same material and were of the same size, the composite beam deflections would be only 25% of those of the non-composite beam and the maximum bending strains (top and bottom surface) would be only 50%.

In composite beams – essentially steel beams supporting the floors in a building or the concrete deck in a bridge – the steel beam is designed to act with a part of the slab in the manner of Figure 9.1(b). For this to happen it is necessary that slip at the interface be prevented. This is normally achieved by the use of devices termed shear connectors. An important aspect of the design of composite beams is therefore the provision of adequate shear connection.

Some indication of the potential benefits achievable by making beams composite with the slab may be obtained from Table 9.1, which is taken from a Swiss publication [1]. Comparing results in the first (non-composite) and last (fully composite) columns, construction depth is reduced by approximately one third, whilst the steel beam weight is almost halved. Clearly if the two components work together, significantly improved performance results; the 'penalty' is the need to provide the necessary shear connection.

Composite action between steel and concrete is not limited to beams. In recent years the benefits of utilizing the potential for composite action

Table 9.1 Comparative beam designs – composite and non-composite (after reference 1)
Self-weight (slab and beam) $w_1 + w_2$
Finishes $w_2 = 1$ kN/m^2
Imposed load $w_2 = 4$ kN/m^2

	Non-composite		Composite	
	Plastic capacity	Elastic capacity	Plastic capacity	
			40% shear connection	60% shear connection
Depth h (mm)	540	500	440	410
Steel beam depth (mm)	400	360	300	270
Steel beam weight (kg/m)	66.3	57.1	42.2	36.1
Number of shear connectors	–	13φ19	10φ19	25φ19
Total deflection (mm)	10	12	24	33
Deflection due to imposed load (mm)	8	3	10	6

between the thin metal sheeting used as permanent formwork to support the concrete slab during casting and the hardened slab have been appreciated [2]; this particular type of composite construction is covered by the Part 4 of BS 5950 [3]. Similarly composite columns – either encased I-sections or filled SHS – offer the potential to carry extremely high loads for relatively small plan areas [4]. Composite action may also be used with advantage in joints [5], in complete frames [4] or in special applications [6, 7]. For building structures an additional advantage is the opportunity to utilize the presence of the concrete as a way of meeting the necessary fire resistance [8].

9.1　Moment capacity of composite beams

Moment capacity of a composite beam is most appropriately calculated using plastic theory in the form of rectangular stress blocks very much in the manner employed for both reinforced concrete and steel. Thus Figure 9.2 illustrates a basic type of cross-section and the associated set of stress blocks for the two cases:

(a) neutral axis in the slab,
(b) neutral axis in the flange.

Considerations of equilibrium of longitudinal forces and internal and external moments give

$$0.45B_e y f_{cu} = A P_y \tag{9.1}$$

$$M_c = A P_y (D_s + D/2 - y/2) \tag{9.2}$$

If the maximum available compression resistance of $0.45B_e D_s f_{cu}$ is less than the tensile resistance of the steel $A p_y$, then the neutral axis falls within the steel section and equations (9.1) and (9.2) should be replaced by

$$0.45B_e D_s f_{cu} + 2A_{sc} p_y = A p_y \tag{9.3}$$

$$M_c = A p_y (D/2 + D_s/2) - 2A_{sc} p_y (h_{sc} - D_s/2) \tag{9.4}$$

In deriving these equations the actual set of stress blocks of Figure 9.2(b) has been replaced with those of Figure 9.3. This is merely a rearrangement for the purposes of simplifying the calculations.

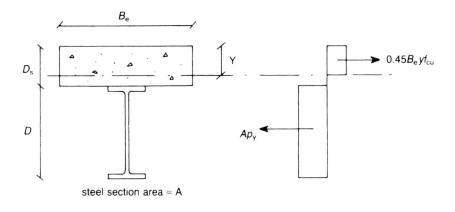

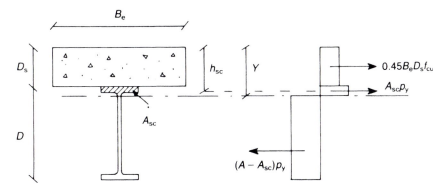

Figure 9.2   Determination of moment capacity: (a) Neutral axis in concrete; (b) neutral axis in steel.

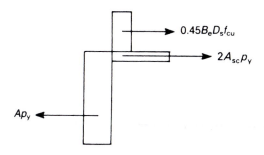

Figure 9.3   Re-arrangement of Figure 9.2(b).

Example 9.1

Calculate the moment capacity of a 457 × 152 × 60 UB of S275 steel when it supports a slab of (a) 200 mm and (b) 100 mm depth. In both cases assume a concrete strength f_{cu} of 30 N/mm² and a slab width $B_e = 1.5$ m.

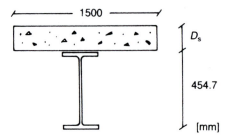

Solution
First check position of neutral axis by comparing tensile resistance of the steel F_T with maximum available compression resistance F_c.

(a) $F_T = Ap_y = 7590 \times 275$
$$= 2087 \text{ kN}$$

$F_c = 0.45B_eD_sf_{cu}$
$$= 0.45 \times 1500 \times 200 \times 30 = 4050 \text{ kN}$$

Since $F_c > F_T$ neutral axis is within slab so use equations (9.1) and (9.2):
$$0.45B_eyf_{cu} = Ap_y$$
$$0.45 \times 1500 \times y \times 30 = 2087$$
$$\text{and} \quad y = 0.103 = 103 \text{ mm}$$
$$M_c = Ap_y(D_s + D/2 - y/2)$$
$$= 2087(200 + 454.7/2 - 103/2) = \underline{784 \text{ kN m}}$$

(b) $F_T = 2087 \text{ kN}$
$F_c = 0.45 \times 1500 \times 100 \times 30 = 2025 \text{ kN}$
Since $F_c < F_T$ neutral axis is in the steel section so use equations (9.3) and (9.4):
$$0.45B_eD_sf_{cu} + 2A_{sc}p_y = Ap_y$$
$$2025 + 2A_{sc} \times 275 = 2087$$
$$A_{sc} = 113 \text{ mm}^2$$

Hence $h_{sc} = 100 + \frac{1}{2}(113/152.9) - 100.4$
$$M_c = Ap_y(D/2 + D_s/2) - A_{sc}p_y(h_{sc} - D_s/2)$$
$$= 2087(454.7/2 + 100/2) - 2 \times 113$$
$$\times 275(101.4 - 100/2)$$
$$= \underline{576 \text{ kN m}}$$

Clearly because the neutral axis for case (b) was so close to the steel–concrete interface, increasing the slab depth to move the neutral axis into the slab in case (a) has comparatively little effect since the tensile force supplied by the steel section can only increase marginally. Virtually the whole of the increase in M_c therefore comes from the increase in lever arm due simply to the increased slab depth.

Assuming the slab depth D_s to have already been decided upon, an estimate for the steel section required to withstand a given moment M may conveniently be obtained by:

1   Estimating the steel area A from

$$A = \frac{2M}{p_y(D_s + D)} \tag{9.5}$$

This assumes the neutral axis to be at the steel/concrete interface.
2   Selecting a suitable steel section based on A.
3   Checking that the neutral axis when using this section will fall within the slab by ensuring that

$$0.45B_cD_sf_{cu} \geqslant Ap_y \tag{9.6}$$

The above approach neglects the small contribution to the cross-section's moment capacity of longitudinal reinforcement in the slab.

Example 9.2

Assuming a slab depth of 130 mm, an effective width $B_c = 1.6$ m and concrete strength corresponding to $f_{cu} = 30$ N/mm², select a suitable steel section to carry a moment of 600 kN m.

Solution

From equation (9.5) estimate $A = \dfrac{2M}{p_y(D_s + D)} = \dfrac{2 \times 600\,000\,000}{275(130 + D)}$

Assuming a 457 UB with $D \approx 460$ mm gives $A = 7396$ mm²

Try 457 × 152 × 60 UB with $A = 7590$ mm²
$$0.45B_cD_sf_{cu} = 0.45 \times 1600 \times 130 \times 30 = 2808 \text{ kN}$$
$$Ap_y = 7396 \times 275 = 2034 \text{ kN}$$

Since $0.45B_cD_sf_{cu} > Ap_y$ neutral axis will be in slab
and $M_c = Ap_y(D_s + D/2 - y/2)$
with $y = Ap_y(0.45B_cf_{cu}) = 2034/(0.45 \times 1600 \times 30) = 94$ mm
∴ $M_c = 2034(130 + 454.7/2 - 96/2) = \underline{631.3 \text{ kN m}}$ (satisfactory).

Appendix B.2.2 of BS 5950: Part 3.1 [9] provides explicit expressions for M_c for the three cases:

1   plastic neutral axis in web;
2   plastic neutral axis in top flange of steel beam;
3   plastic neutral axis in slab.

These are expressed in terms of the force components (products of stress times the area over which it acts) for the different parts of the cross-section:

$$R_s = Ap_y$$ Resistance of steel beam
$$R_c = 0.45f_{cu}B_eD_s$$ Resistance of concrete flange
$$R_f = BTP_y$$ Resistance of steel flange
$$R_w = R_s - 2R_f$$ Resistance of overall web depth
$$R_v = dtp_y$$ Resistance of clear web depth

Thus for case (1) equation (9.4) has been rewritten as

$$M_c = M_s + R_c\frac{(D + D_s)}{2} - \frac{R^2_c}{R_v}\frac{d}{4} \tag{9.7}$$

in which M_s = plastic moment capacity of the steel section.
   Should the neutral axis fall within the upper flange of the steel beam, then equation (9.7) must be replaced by

$$M_c = R_s\frac{D}{2} + R_c\frac{D_s}{2} - \frac{(R_s - R_c)^2}{R_f}\frac{T}{4} \tag{9.8}$$

In practice since this case implies that R_s and R_c will be approximately equal, the last term will normally be small and may reasonably be ignored.
   For case (3) equation (9.2) becomes

$$M_c = R_s\left[D/2 + D_s - \frac{R_s}{R_c}\frac{D_s}{2}\right] \tag{9.9}$$

Example 9.3

Rework Example 9.1 using the Part 3.1 format of equations (9.7) and (9.8).

Solution
(a) $R_s = Ap_y = 7590 \times 275 = 2087$ kN
    $R_c = 0.45f_{cu}B_eD_s = 0.45 \times 30 \times 1500 \times 200 = 4050$ kN

Since $R_c > R_s$, neutral axis is within slab, so use equation (9.8):

$$M_c = R_s \left[D/2 + D_s - \frac{R_s}{R_c} \frac{D_s}{2} \right]$$

$$= 2087 \left[\frac{454.7}{2} + 200 - \frac{2087}{4050} \times \frac{200}{2} \right]$$

$$= \underline{784 \text{ kN m}}$$

(b) $R_s = 2087$ kN
$R_c = 0.45 \times 30 \times 1500 \times 100 = 2025$ kN.

Since $R_c < R_s$ neutral axis is in the steel section so use equation (9.8):

$$M_c = R_s \frac{D}{2} + R_c \frac{D_s}{2} - \frac{(R_s - R_c)^2}{R_f} \frac{T}{4}$$

and $R_f = BTp_y = 152.9 \times 13.3 \times 275 = 559$ kN

$$\therefore M_c = 2087 \frac{454.7}{2} + 2025 \frac{100}{2} - \frac{(2087 - 2025)}{559} \frac{13.3}{4}$$

$$= 10^6 (474.5 + 101.2 - 0) = \underline{575.7 \text{ kN m}}$$

For case (a) the result is identical with that obtained using the original equation (9.2). A small difference (less than 1%) exists between the two treatments when the neutral axis is in the steel section, largely because of the treatment of terms in the formulae that make only a small contribution.

For both cases the value of M_c should be compared with that of 352 kN m for the steel section acting alone. Thus the increases in M_c are 123% and 64% respectively.

In deriving all of the equations of this section it has been assumed that the plate elements of the steel section are of such proportions that the composite beam may be classified as a 'plastic' cross-section. Because the strain profiles in a composite beam will be different from those of bare steel sections it does not automatically follow that the width/thickness limits of *Table 11* of Part 1 will be appropriate. For simply supported beams under positive moment, however, the top flange will be supported against local buckling by the slab, the bottom flange will be in tension and thus only the web presents a potential problem. The required d/t limit is provided in *Table 11* as

$$d/t \leqslant \frac{80\epsilon}{1 + r_1} \text{ but } \geqslant 40\epsilon \qquad (9.10)$$

in which $r_1 = F_c/dtp_{yw}$

As the value of r_1 increases from -1 (corresponding to the neutral axis being located at the interface of the steel beam and the slab for which F_c is tensile and equal to $-dtp_{yw}$) so progressively more of the web will be in compression. However, even for $r = 0$ (neutral axis at mid-depth of the steel section) the d/t limits from (9.10) for S275 and S355 material are 80 and 70 respectively. Since no UB sections have d/t values above 55, it follows that the positive moment capacity of composite beams having any slab proportions which use these sections as the steel part may be determined using the methods of this section, in particular equations (9.7)–(9.9).

Thus for simply supported beams local buckling and cross-section classification will not normally be an issue providing the steel section is a conventional hot-rolled UB or UC. If a more slender fabricated plate girder is employed such that the above limits are exceeded, then the design approach of reference [8] is to neglect the contribution of part of the web (rather in the manner described previously for bare steel members containing slender plate elements in Section 5.3.1). Local buckling is, however, likely to become a major design consideration in negative moment regions, e.g. the support regions of continuous composite beams; readers wishing to learn something of this should consult more specialized texts on composite construction [4, 10, 11].

9.2 Shear connection

For the behaviour assumed in the previous section when determining M_c to be valid, slip at the steel–concrete interface must be prevented. Much the most widely used type of shear connector is the headed stud illustrated in Figure 9.4. Although available in various sizes as indicated, 19 mm studs of 75 mm height account for most of the applications in buildings.

Studs may be welded either in the shop or on site using a special form of 'gun'. A particularly simple type of bend test in which sample studs are either hit with a hammer or bent over using a scaffold tube is normally all that is required to check the integrity of the welding.

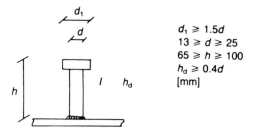

$$d_1 \geq 1.5d$$
$$13 \geq d \geq 25$$
$$65 \geq h \geq 100$$
$$h_d \geq 0.4d$$
[mm]

Figure 9.4 Headed shear stud.

For design purposes the only property that is required is slip load Q_k; this is typically found from a push-out test using an arrangement of the type shown in Figure 9.5. Although standardized procedures for conducting push-out tests are available (reference [9] refers the reader to the composite part of the bridge code [12]), leading manufacturers normally provide suitable values for Q_k based on their own tests. Table 9.2 gives design values for the static strength of studs in plain concrete [8]. For lightweight concrete 90% values of these should be used.

Design calculations for shear connectors may be based on the simple requirement that a flexural failure is achieved, i.e. the degree of shear connection must be sufficient to prevent a shear failure, at a moment at least equal to M_c as given by equation (9.2) or (9.4). Although cases can be made for various arrangements, providing heavy point loads are not present, it is normally quite sufficient to employ a uniform spacing. Some empirical limits on spacing are also necessary so as to prevent uplift of the

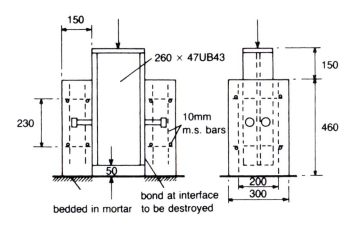

Figure 9.5 Push-out test.

Table 9.2 Design strengths of shear connectors in normal weight concrete slabs to be used for regions of positive moment (taken as 0.8 times the characteristic resistances of *Table 5* of ref [9])

Dimensions of stud (mm)		Characteristic strength of concrete (N/mm²)			
d	h	25	30	35	40
25	100	117	123	129	134
22	100	95	101	106	111
19	100	76	80	83	87
19	75	66	70	73	77
16	75	56	59	62	66

slab, to ensure a smooth flow of shear into the concrete etc.; these are listed in *Cl. 5.4.8* of BS 5950: Part 3.1 [9].

Example 9.4

Determine the number of 19 mm by 75 mm shear studs required for case (b) of Example 9.1.

Solution
From Table 9.2 design strength per stud (assuming normal weight concrete) = 70 kN
Since longitudinal force that needs to be transferred = F_c of 2025 kN number of studs required = 2025/70 = 29.
∴ use 30 connectors, arranged as 15 pairs spaced uniformly.

9.3   Other design considerations

9.3.1   Steel beam

It is usual to assume that all of the vertical shear in the composite beam is carried by the steel section alone. Thus the shear capacity will be (as before)

$$V_b = 0.6p_y dt \tag{9.11}$$

Should the applied shear exceed 50% of this figure, then *Cl. 5.3.4* provides guidance on the necessary reduction in moment capacity.

9.3.2   Concrete slab

Thus far it has been assumed that the extent of the slab of a composite beam is defined (dimension B_e in Figure 9.2). In reality the slab will be continuous over a number of beams as shown in Figure 9.6.

It is therefore necessary to identify that part which may reasonably be assumed to act with the steel section as a composite beam. (A similar problem is encountered in concrete construction when the ribs supporting a slab are to be designed as tee-beams.)

The typical variation of longitudinal stress in the concrete flange sketched in Figure 9.6 suggests the use of an effective breadth of slab, defined in such a way that the application of simple bending theory to the effective cross-section will give broadly the same result as would be obtained by considering the true behaviour of the actual cross-section. Much has been written [4] about effective breadths, indicating that values

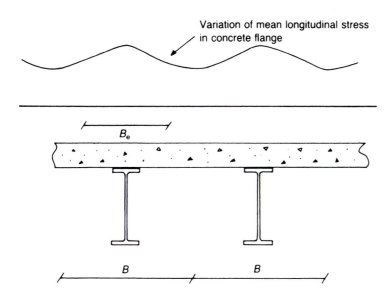

Figure 9.6   Effective breadth of concrete flange.

depend in a complex fashion upon the ratio of beam spacing to span, the form of the applied loading, the support conditions and the load level. However, BS 5950: Part 3.1 simply requires that B_e/L should not exceed 0.25, with the effective slab material being symmetrically disposed.

Failure of the slab due to the longitudinal shear transmitted by the shear connectors within the region of the effective breadth may occur unless sufficient transverse reinforcement of the type shown in Figure 9.7 is provided. The shear v to be resisted per unit length is simply the force developed by the shear connectors, viz.

$$v = N(0.8Q_k)/s \tag{9.12}$$

in which

$$N = \text{number of studs in group}$$
$$0.8Q_k = \text{design strength of stud}$$
$$s = \text{longitudinal stud spacing.}$$

Sufficient resistance must be provided to resist this force at every potential shear failure surface within the slab. Thus v must be less than the shear resistance v_r, given by the lesser of [9]:

$$v_r = 0.7A_{sv}f_y + 0.03A_{cv}f_{cu} \tag{9.13a}$$

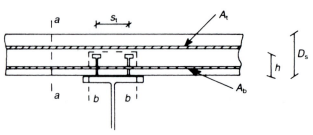

shear surface	length	A_{sv}
aa	D_s	$A_b + A_t$
bb	$2h + s_t{}^*$	$2A_b$

* 2h for either a single row or staggered studs

Figure 9.7   Shear failure and transverse reinforcement.

or

$$v_r = 0.8A_{cv}\sqrt{f_{cu}} \qquad\qquad (9.13b)$$

in which

A_{sv} = cross-sectional area per unit length of reinforcement
A_{cv} = mean cross-sectional area per unit length of concrete shear
  surface under consideration.

Figure 9.7 indicates the two possible failure surfaces – aa and bb – as well as giving expressions for A_{sv} for both. Other cases such as haunched slabs and the presence of metal sheeting (see Section 9.6) are covered in reference [8].

The background theory leading to the development of equation (9.13) is provided in reference [4]. Design of the slab will already have fixed the arrangement of top reinforcement and thus the value of A_t. Thus design for longitudinal shear will normally reduce to a check on the need for bottom reinforcement using equation (9.13a) and shear surface bb. In many cases the provision of bottom reinforcement will be found to be unnecessary.

9.4   Partial shear connection

In cases where the sizes of the slab and the steel beam are decided upon on the basis of considerations other than their combined strength as a

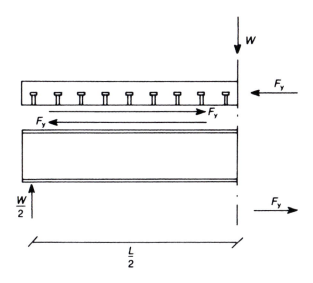

Figure 9.8   Free body diagram for one half span.

composite beam, it may be advantageous not to have to provide sufficient shear connection to produce full interaction, since a lesser moment capacity will be adequate. This may be achieved by reducing the number of shear connectors. However, if too few are provided the degree of slip that will occur, even at comparatively low loads, will be so great that those connectors present will shear off and the strength of the composite section will simply correspond to that of the steel member.

To understand the basis for partial interaction design it is first necessary to consider the effect of varying the degree of shear connection on load-carrying capacity. Referring to the basic free body diagram for the left-hand part of a centrally loaded composite beam shown in Figure 9.8, the degree of shear connection r is defined by

$$r = NQ_k/R_s \tag{9.14}$$

in which

N = number of shear connectors
Q_k = connector strength (as determined from push-out tests)

Based on a sophisticated analysis [13] that allows for the presence of the shear connectors by direct use of their load-slip behaviour, results of the type given as Figure 9.9 may be derived for particular cross-sections. Three types of behaviour may be observed:

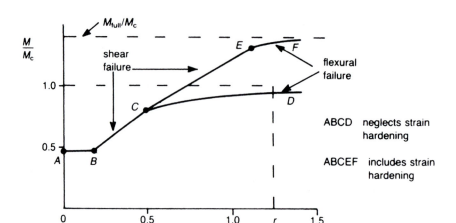

Figure 9.9   Effect of degree of shear connection on moment capacity.

AB   Insufficient shear connection for any composite action.

BC or CE   Some composite action but studs shear off before failure of steel or concrete.

CD or EF   Flexural failure that, providing strain hardening in the steel section is considered (see Section 1.2), will ensure that a moment at least equal to M_c will be achieved.

The alternative simplified analysis leading to curve ABCD of Figure 9.9 is included here merely to illustrate the need to consider strain-hardening in this particular application if meaningful results are to be obtained. It is, of course, exhibited by normal structural steels.

Noting that shear failure is less predictable and is also likely to be more sudden than flexural failure, BS 5950: Part 3.1 requires that design be based on flexural failure. To achieve this Figure 9.9 shows that r should not be less than about 1.25. Taking the design strength Q_d of a stud as $0.8Q_k$, assuming that R_f is resisted equally by N shear connectors so that $R_f = NQ_d$ and substituting in equation (9.14) gives

$$r = 1.25 \qquad (9.15)$$

Thus beams designed on the basis of $r = 1.25$ should fail in flexure at moments not less than M_c calculated from equation (9.4). This is referred to as 80% design; the number of shear connectors required is N_p.

When fewer shear connectors than N_p are used the relationship between moment capacity M and the actual number N will be as shown by curve AB in Figure 9.10. Low levels of shear connection should be avoided for

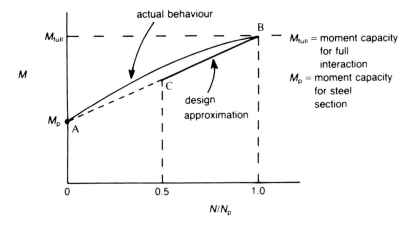

Figure 9.10   Design approach for incomplete shear connection.

the reasons already mentioned and a lower limit on N/N_p is common [9, 14]. This leads to the design rule based on line CB of

$$M = M_p + \frac{N}{N_p}(M_{\text{full}} - M_p)$$ (9.16)

for $\dfrac{N}{N_p} \geqslant 0.5$

Part 3.1 of BS 5950 [9] presents this concept rather differently, simply requiring that the actual moment capacity of a section with partial shear connection be determined using a reduced value for the force in the concrete slab equal to $N_a Q_p$, where N_a is the actual number of connectors used. This then leads to a reduced depth for the concrete stress block (y in Figure 9.2a). A lower limit of $0.4N_p$ is placed on N_a for spans up to 10 m. Because of concern over the absolute magnitude of slips in longer partially connected beams, a linear increase in this limit up to 100% at $L = 16$ m is given.

Example 9.5

For the beam of case (b) of Example 9.1 consider the effect of reducing the degree of shear connection by using (a) 24 studs; (b) 20 studs.

Solution

(a) $M_c = M_p + \dfrac{N}{N_p}(M_{full} - M_p)$

$= 275 \times 1\,280\,000 \times 10^{-6} + \dfrac{24}{30}(570 - 275 \times 1\,280\,000 \times 10^{-6})$

$= 352 + 0.8(570 - 352) = \underline{526\ kN\ m}$

(b) $M_c = 352 + 0.67(570 - 352) = \underline{498\ kN\ m}$

9.5   Serviceability considerations

Thus far in this chapter attention has been focused solely upon the ultimate limit state. Composite beams do, however, require rather more attention under serviceability conditions than bare steel beams, both in terms of deflections and in terms of stresses. When used in situations in which dynamic loading is present, e.g. bridges, their fatigue performance will often be a major factor in their design [10].

Deflections and stresses under static loading may be calculated by elastic analysis using a transformed sections approach assuming full interaction [4]. The method of construction (although it does not affect ultimate load carrying capacity) is of importance due to the two different cases.

1 *Propped construction.* The steel beams are supported on props during the wet concrete stage; the whole of the load (dead and imposed) is therefore carried by the composite section.
2 *Unpropped construction.* The steel beams must support the wet concrete during the construction phase, with the composite section being available to support the imposed loads.

Of the two, unpropped construction is the more usual in buildings, simply because it avoids the cost (in both money and time) of the propping operation. Usually, only if serviceability deflections were found to be unacceptably large would propping be considered. In addition shrinkage and creep of concrete will also contribute to service deflections, so three types of loading should be considered:

1 that carried by the steel alone;
2 long-term loading on the composite section;
3 short-term loading on the composite section.

Just as for reinforced concrete sections the elastic analysis assumes that plane sections remain plane and that the concrete cannot carry any tensile stress. This leads to the representation of Figure 9.11. The modular ratio should be obtained from:

for short-term loading $\alpha_s = E_s/E_c$
for long-term loading $\alpha_L = E_s/k_c E_c$

in which

k_c = creep factor for concrete
E_s, E_c = modulus of elasticity for steel and concrete respectively.

Table 1 of BS 5950: Part 3.1 provides values for α_s and α_L for both short-term and long-term loading. Providing these are used it is not necessary to make any additional allowances for the effects of creep and shrinkage [15] on deflections. Moreover, loading may normally be assumed to comprise two thirds short term and one third long term, leading to an α_e value of 18 for normal weight concrete. For deflection calculations the gross value of second moment of area for the uncracked section I_g should be used; this is given as

$$I_g = I_x + \frac{B_e D_s^3}{12\alpha_e} + \frac{AB_e D_s (D + D_s)^2}{4[A\alpha_e + B_e D_s]} \qquad (9.17)$$

The possibility that deflections under service loads might become too large due to irreversible material effects may be eliminated by ensuring that the maximum stresses in the steel and the concrete do not exceed p_y and $0.50f_{cu}$ respectively. Elastic calculations using an elastic section modulus determined from either I_g if the elastic neutral axis is in the steel section or the cracked section value I_p if the elastic neutral axis is in the concrete slab should be used. The appropriate formulae are given in Table 9.3. The exact process to be followed in either case is illustrated by Example 9.6.

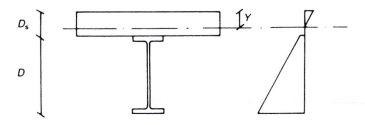

Figure 9.11   Analysis for serviceability conditions.

Table 9.3 Properties for use in serviceability calculations for beams with solid slabs

	Elastic neutral axis in steel beam	Elastic neutral axis in concrete slab
governing condition	$A \geqslant \dfrac{D_s^2 B_e}{D\alpha_e}$	$A < \dfrac{D_s^2 B_e}{D\alpha_e}$
section modulus, concrete slab	$Z_g = I_g \alpha_e / y_g$	$Z_p = I_p \alpha/y_e$
section modulus, steel flange	$Z_s = I_g/(D - y_g)$	$Z_s = I_p/(D - y_e)$
depth of neutral axis below top surface of slab	$y_g = \dfrac{A\alpha_c(D + 2D_s) + B_e D_s^2}{2[A\alpha_e + B_c D_s]}$	$y_e = \dfrac{D}{1 + \left[1 + \dfrac{B_e}{A\alpha_e}(d + 2D_s)\right]^{\frac{1}{2}}}$
second moment of area	$I_g = I_x + \dfrac{B_e D_s^3}{12\alpha_e} + \dfrac{AB_e D_s(D + D_s)^2}{4[A\alpha_e + B_e D_s]}$	$I_p = I_x + \dfrac{B_e y_e^3}{3\alpha_e} + A\left(\dfrac{D}{2} + D_s - y_e\right)^2$

The deflections of beams designed on the basis of partial interaction may be approximated, using reference [3]:

$$\delta = \delta_f + \tfrac{1}{2}(\delta_s - \delta_f)\left(1 - \frac{N}{N_p}\right) \tag{9.18}$$

in which

δ_f = deflection assuming full interaction
δ_s = deflection for steel section acting alone.

An alternative arrangement of equation (9.18) is possible as

$$\delta = \delta_f + \tfrac{1}{2}\delta_f\left(\frac{I_c}{I_s} - 1\right)\left(1 - \frac{N}{N_p}\right) \tag{9.19}$$

in which I_s, I_c = second moments of area of steel and composite sections respectively.

As an alternative to the actual calculation of deflections, a more rapid check for acceptability may be made using span–depth charts [16]. This concept, which is well established in reinforced concrete construction [15], simply requires that the actual ratio of clear span to overall depth be kept below a certain limit. Basic values have been provided that ensure that deflections will not exceed span/250 for different:

1   concrete cube strength and density;
2   slab depth/steel section depth (D_s/D);
3   slab area/steel section area (A_c/A);
4   grade of steel (p_y).

Modifications are possible to allow for:

1   unpropped construction;
2   long spans (>10 m);
3   concrete strength in excess of 20 N/mm²;
4   lightweight concrete;
5   partial interaction design.

Example 9.6

Assuming the beam of case (b) of Example 9.1 to be simply supported over a span of 9 m at a spacing of 4.5 m and to be supporting a total imposed working load of 7.5 kN/m² investigate the serviceability deflections for full and partial interaction design.

Solution

$$\delta_{max} = \frac{5wL^4}{384EI}$$

From equation (9.17)

$$I_g = I_x + \frac{B_eD_s^3}{12\alpha_e} + \frac{AB_eD_s(D + D_s)^2}{4[A\alpha_e + B_eD_s]}$$

$$= 255\,000\,000 + \frac{1500 \times 100^2}{12 \times 18} + \frac{7590 \times 1500 \times 100(454.7 + 100^2)}{4[7590 \times 18 + 1500 \times 100]}$$

$$= 10^6[255 + 7 + 306] = 568 \times 10^6 \text{ mm}^4$$
$$w = 7.5 \times 4.5 = 33.8 \text{ kN m}$$
$$= 33.8 \text{ N/mm}$$

$$\therefore \delta = \frac{5 \times 33.8 \times 9000^4}{384 \times 205\,000 \times 568 \times 10^6} = \underline{24.8 \text{ mm}}$$

This is span/363, which is small.

Assuming that the dead load is 2.5 kN/m² and the use of unpropped construction so that this had to be resisted by the steel beam alone would give

$$\delta = \frac{5 \times 11.3 \times 9000^4}{384 \times 205\,000 \times 255 \times 10^6} = \underline{18.5 \text{ mm}}$$

or span/486, which is also within normal limits.

If a partial interaction design is to be used, equation (9.18) gives

$$\delta = \delta_f + \tfrac{1}{2}(\delta_s - \delta_f)\left(1 - \frac{N}{N_p}\right)$$

in which $\delta_s = 24.8\,\dfrac{568}{255} = 55.1 \text{ mm}$.

For $N/N_p = 0.67$ (20 studs)
$$\delta = 24.8 + \tfrac{1}{2}(55.1 - 24.8)(1 - 0.67)$$
$$= 24.8 + 5.0 = \underline{29.8 \text{ mm}}$$

This corresponds to span/302 and would normally be considered acceptable [16].

Using method A of reference [15] assuming full interaction,
$$D_s/D = 100/454.7 \qquad = 0.22$$
$$A_c/A = 1500 \times 100/7590 = 19.76$$

From *Figure A2* limiting $R = 21.6$.
For selected beam $R = 9000/4454.7 + 100) = 16.2$.

Long span fascia girder in a new stand at Ibrox.

Since this is less than the limiting value design is OK. (It should be noted that the chart of reference [15] assumes a limiting deflection of span/215 and the use of Grade 20 concrete.)

9.6   Use of metal sheeting

Since it is necessary to support the wet concrete of the slabe during construction, it is clearly likely to be advantageous if the support system can be left in place and made to contribute structurally to the final arrangement. One of the most significant contributions to the rapid growth in the use of steel for multistorey construction in Britain during the 1980s has been the utilization of floor arrangements of the type illustrated in Figure 9.12. This uses profiled steel sheets of around 50 mm depth and 0.90 mm material thickness, typical examples of which are shown in Figure 9.13, to span between beams and to act as both permanent formwork and tension reinforcement for the slab. (The behaviour and design of composite slabs is considered in Section 9.7.)

Thus two forms of composite action are now being employed: between the slab and the sheeting to span transversely and between the slab and the steel section to span longitudinally. Shear studs may be site welded through the sheets, which are initially secured in place with steel pins fired

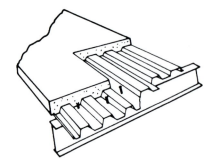

Figure 9.12   Composite beam incorporating profiled sheeting.

a)   Re-entrant   b) Trapezoidal

Figure 9.13   Typical profiles–various depths, various forms of indentation and various types of stiffening are used.

through the sheet and the beam top flange. Thus the sheeting also provides both a temporary working platform and a shield to those working lower down on the building and thereby makes a major contribution towards improving productivity on site.

In terms of composite beam design, the process remains essentially similar to that already described, with the following provisos.

1   Push-out tests on specimens that incorporate sheeting give lower slip strengths than those in plain concrete. A lower value of Q_d should therefore be used and *Cl. 5.4.7* gives reduction factors to be applied to the basic design strengths listed in Table 9.2 for different arrangements of studs, profile geometries, etc.
2   Only that depth of concrete above the top surface of the sheeting may be assumed to constitute the slab depth $(D_s - D_p)$.

In order to assist designers (as well as to market their product) most sheeting manufacturers have contributed to the compendium of design tables produced by the SCI [17]. This obviates the need for detailed calculation.

For the case in which the sheeting is arranged to span parallel to the beams, it is normally possible to use the full slab depth as h_c and to design the cross-section according to Sections 9.1–5.

9.7   Composite slabs

The concrete slab, acting in conjunction with the sheeting in Figure 9.13, behaves rather in the manner of an under-reinforced concrete beam spanning between the parallel steel beams. The sheeting provides the reinforcement. To do this there must be sufficient bond developed between the surfaces in contact. In normal reinforced concrete this is achieved by roughening the surface of the reinforcing bars during manufacture. A similar technique is used for many of the profiles intended for use in composite slabs. For the trapezoidal type profile of Figure 9.13 various types of indentation may be formed in the webs and/or flanges during the forming process. These provide a mechanical means of resisting slip, acting as a series of keys into the concrete. On the other hand re-entrant profiles are of such a shape that they are forced to push against the concrete when the slab is loaded, thereby resisting slip through friction, although improved bond may be obtained by the use of indentations as well.

However, with those profiles presently available and for the range of spans, slab depths, load conditions etc. required in practice, it has not been possible to eliminate the shear-bond failure as a possible mode – as may be done for composite beams for example by a correct choice of r (Figure 9.9). Thus the part of BS 5950 that deals explicitly with composite slabs [3] includes a procedure for determining failure loads based on shear-bond behaviour. For long slabs this mode will cease to govern, in which case the slab may be designed for flexure as a normal reinforced concrete beam [3].

For many composite slabs, however, the ability to support the wet concrete without undue deflection over spans of the order of 3 m is likely to be the governing criterion. Methods for assessing the behaviour of the decking when acting on its own are provided in reference [2]. Since that approach tends to be rather conservative when compared with the performance observed in full-scale tests, other methods [18] tend to be favoured by manufacturers as the basis for their own design data.

Because of the extensive use of metal decking in composite construction, the formulae for moment capacity, second moment of area, section modulus etc. provided in BS 5950: Part 3.1 are presented in terms of a slab depth D_s that includes a deck of depth D_p. Thus the equivalents of equations (9.7), (9.8), (9.13), Table 9.3, etc. are actually the only formulae given in the code.

Further developments have seen special types of both metal making and beam section produced as the basis for a number of composite floor arrangements designed to optimize structural performance under both normal and fire conditions [19]. Full design information is available in a series of SCI publications [20–23].

References

1 Bucheli, P. and Crisinel, M. (1982) *Poutre Mixte dans le Bâtiment*, Centre Suisse de la Construction Metallique, Zurich.
2 Evans, H.R. and Wright, H.D. (1988) Steel–concrete composite flooring deck structures, in *Steel–Concrete Composite Structures: Stability and Strength* (ed. R. Narayanan), Elsevier Applied Science Publishers, pp. 21–52.
3 British Standards Institution (1994) BS 5950: Part 4, *Structural Use of Steel in Building. Code of Practice for Design of Composite Slabs with Profiled Steel Sheeting*, London.
4 Johnson, R.P. (1994) *Composite Structures of Steel and Concrete*, Vol. 1, *Beams, Columns, Frames and Applications in Building*, Blackwell Scientific Publications, London.
5 Zandonini, R. (1989) Semi-rigid composite joints, in *Connections: Stability and Strength* (ed. R. Narayanan), Elsevier Applied Science Publishers, pp. 63–120.
6 Liauw, T.C. (1988) Steel frames with concrete infills, in *Steel–Concrete Composite Structures: Stability and Strength* (ed. R. Narayanan), Elsevier Applied Science Publishers, pp. 115–62.
7 Narayanan, R., Wright, H.D., Evans, H.R. and Francis, R.W. (1987) Double-skin composite construction for submerged tube tunnels, *Steel Construction Today*, **6**, 185–90.
8 British Standards Institution (1990) BS 5950: Part 8, *The Structural Use of Steelwork in Building Code of Practice for Fire Resistant Design*, London.
9 British Standards Institution (1990) BS 5950: Part 3.1, *The Structural Use of Steelwork in Building, Code of Practice for Design in Composite Construction. Design of Simple and Continuous Composite Beams*, London.
10 Johnson, R.P. and Buckby, R.J. (1986) *Composite Structures of Steel and Concrete*, Vol. 2, 2nd edn, Blackwell Scientific Publications, Oxford.
11 Bradford, M.A. and Johnson, R.P. (1987) Inelastic buckling of composite bridge girders near internal supports, *Proceedings Institution of Civil Engineers*, **83** (Part 2), 143–59.
12 British Standards Institution (1978) BS 5400: Part 5, *Steel Concrete and Composite Bridges*, London.
13 Yam, L.C.P. and Chapman, J.C. (1968) The inelastic behaviour of simply supported composite beams of steel and concrete, *Proceedings Institution of Civil Engineers*, **41**, 651–84.
14 Kulak, G., Adams, P.F. and Gilmor, M.I. (1990) *Limit States Design in Structural Steel*, 4th edn, Canadian Institute of Steel Construction.
15 Kong, F.K. and Evans, R.H. (1980) *Reinforced and Prestressed Concrete*, 2nd edn, Nelson, Surrey.
16 Johnson, R.P. and Smith, D.G.E. (1975) Design rules for the control of deflections in composite beams, *The Structural Engineer*, **53**(9), 367–76.
17 Lawson, R.M. (1989) *Design of Composite Slabs and Beams with Steel Decking*, The Steel Construction Institute, P. 055.
18 Bryan, E.R. and Leach, P. (1984) *Design of Profiled Sheeting as Permanent Formwork*, CIRIA Technical Note 116.
19 Lawson, R.M. et al (1999) 'Slimflor' and 'Slimdeck' Construction: European Developments, *The Structural Engineer*, Vol. 77, No. 8, April, pp. 22–50.
20 Mullett, D.L. (1991) 'Slim-floor Construction', The Steel Construction Institute, P. 100.
21 Mullett, D.L. and Lawson, R.M. (1993) 'Slim-floor Construction using Deep Decking', The Steel Construction Institute, P. 127.

22 Lawson, R.M., Mullett, D.L. and Rackham, J.W. (1997) 'Design at Asymetric "Slimfloor" Beams using Deep Composite Decking", The Steel Construction Institute, P. 175.
23 Mullett, D.L. (1999) 'Design of RHS "Slimfloor" Edge Beams', The Steel Construction Institute, P. 169.

Chapter 10

Frames

The previous chapters – especially 3–6 – have been concerned with the behaviour of individual elements assuming both the loading and support conditions to be known. Apart from the treatment of end-condition effects for struts in Section 4.3 and some general comments on the development of end moments for beam columns in Chapter 6, the question of inter-action between components has not progressed beyond the introductory discussion of framing types of Section 2.1. This was deliberate as it is necessary to possess a sound understanding of the response of the different types of structural element in clearly defined situations before attempting to consider them as parts of structures.

Section 2.1 did, however, draw out the important distinction between the two main forms of framing considered by BS 5950: Part 1: 'simple construction' and 'continuous construction'. These points will be developed further in this chapter as part of a wide-ranging discussion of the behaviour of frames. Detailed points relating to the design of components in frames conceived according to the principles of continuous construction are covered in Chapter 11. Useful guidance on the overall analysis of frames, i.e. determination of the distribution of internal forces and moments produced by the applied loading(s) may be found in the SCI's Guide on how best to utilize standard computer package programs so as to properly represent a number of important practical features [1].

10.1 Simple construction

Ideally frames designed on the basis of simple construction should utilize joints between members that possess negligible rotational stiffness and thus are incapable of transmitting moments around the structure. This would then permit all members to be designed essentially in isolation, either as axially loaded ties or struts or as simply supported beams. Figure 10.1 illustrates the concept for some simple examples.

Real structural joints, whether between beams and columns (Section 8.1) or in trusses (Section 8.8), will not conform exactly to this ideal. In addition it may well be more practical to 'run through' certain members to take advantage of stock lengths and to avoid unnecessary joints, e.g. the chords of the truss of Figure 10.1(a) and the columns of the frame of Figure 10.1(b) would not normally be broken at every intersection. Thus some judgement is necessary in deciding that a particular configuration may safely be treated as 'simple construction' and some corrections and empirical factors will appear in the basic design rules so as to make approximate allowances for the differences between assumed and actual behaviour. Some illustrations of this have already been given, e.g. the treatment of load eccentricity for angle tension members (Section 3.3), the differences between theoretical and design values of effective length factors for struts (Section 4.3.1) etc.

10.1.1 Trusses

The types of joints actually used in steel trusses, e.g. Figure 8.17 and 8.18, will not, of course provide the 'perfectly pinned connections' usually assumed when determining the distribution of internal member forces. However, the triangulated nature of the framing will mean that the principal method of resisting the external loads will be by the development of a set of tensile and compressive member forces; bending effects will usually be of much less significance. Thus, with the exception of major structures, e.g. very long span roof trusses several metres deep of the sort used for the roof of the turbine hall in a power station, it is customary to design trusses and lattice girders of the type shown in Figure 10.2 as if they were pin-jointed. To assist with this, *Cl. 4.10* of BS 5950: Part 1 provides a set of simplified design rules.

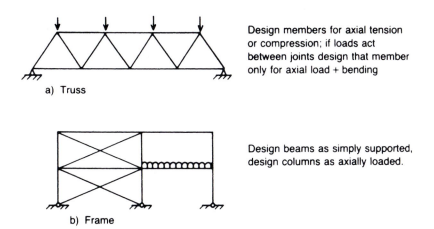

a) Truss

Design members for axial tension or compression; if loads act between joints design that member only for axial load + bending

b) Frame

Design beams as simply supported, design columns as axially loaded.

Figure 10.1 Idealized structural framing arrangements.

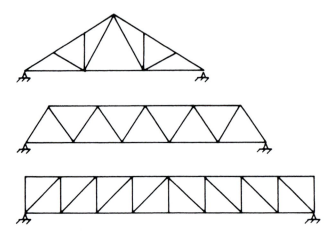

Figure 10.2   Trusses and lattice girders.

Essentially these state that bending effects due to joint rigidity may be neglected and the truss designed for a set of axial member forces determined from an analysis that assumes pin-joints providing:

1   Both in-plane and out-of-plane behaviour must be considered, particularly with respect to buckling.
2   Effective lengths may be determined taking into account restraint from adjacent members.
3   Bending moments due to point loads applied between joints may be taken as $WL/6$, where L is the distance between joints.

When designing roof trusses it is particularly important to identify those members which might suffer compression under certain types of applied loading even though such forces might not be the absolute maximum values.

10.1.2   Rectangular frames

By far the most widely employed arrangement for building frames in the UK is a multistorey steel frame, designed to support gravity loading, acting in conjunction with either diagonally braced bays, cores or shear walls that are assumed to resist the whole of the lateral loading. Figure 10.3 illustrates the concept. Proper structural connection between the steel frame and the concrete core is necessary in order that the frame can transfer horizontal loads into the core and thus 'lean' against it. An example of such a connection is given in Figure 10.4; it is important that the particular

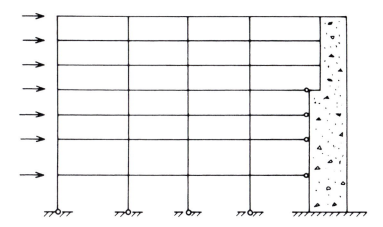

Figure 10.3   Lateral support for simply constructed steel frame from concrete core.

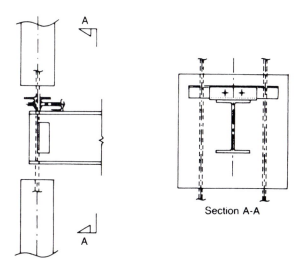

Section A-A

Figure 10.4   Detail for attachment of steel beam to concrete core.

arrangements adopted provide some degree of dimensional tolerance to assist with positioning of the end of the incoming beam.

Since there is no frame action present, externally applied loads are usually distributed in a simple statical manner based upon areas. Such a process is facilitated by the normal assumption that floor slabs span in one direction (one-way spanning) so that the load path is

floor slab → secondary beams →
primary beams → columns → foundations

Figure 10.5 illustrates the concept of allocating loads based on floor areas as well as the onward transfer. This then leads to the process of accumulating column loads both from incoming beams and from the floor levels above shown in Figure 10.6.

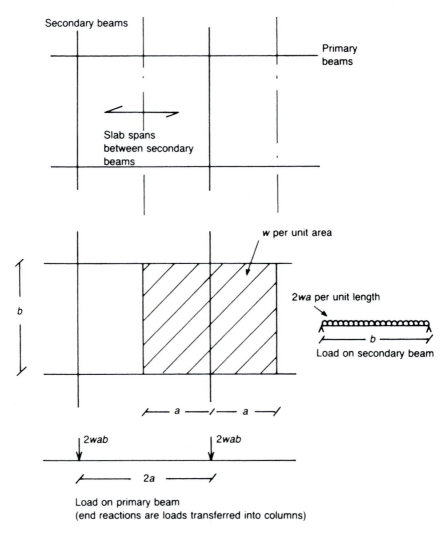

Figure 10.5   Allocation of floor loads.

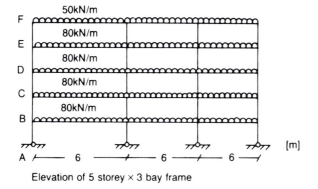

Elevation of 5 storey × 3 bay frame

Storey	Loading (kN)	Self-wt at column assumed	Total load (kN)
5	$\dfrac{200}{6}$	3	209
4	$\dfrac{320}{6}$	3	518
3	$\dfrac{320}{6}$	5	839
2	$\dfrac{320}{6}$	5	1170
1	$\dfrac{320}{6}$	5	1501

Figure 10.6 Accumulation of load for a corner column assuming all floor loads carried by primary beams.

10.1.3   Load cases

All parts of a structure should be designed to withstand the most severe loading that they can reasonably be expected to receive during the life of the structure. As discussed in Chapter 2, this requires judgements to be made not just on load levels, e.g. the likely magnitude of floor loading for office buildings, but also on frequency of occurrence, e.g. the number of times that an 80 m.p.h. wind will occur over a 50-year period. For certain classes of structure it will be necessary to consider a load spectrum, i.e. the mix of severity and frequency, and perhaps also to use this in a dynamic fashion, e.g. to assess earthquake loading for structures in seismically active regions. Fortunately for virtually all building structures in the UK it is sufficient merely to work in terms of static loads and to refer to BS 6399 [2] for detailed information.

Taking these loadings in conjunction with the requirements of BS 5950: Part 1 leads to the three basic cases:

1	dead load + imposed load	($\gamma_D = 1.4$, $\gamma_L = 1.6$)
2	dead load + wind load	($\gamma_D = 1.4$ or 1.0, $\gamma_w = 1.4$)
3	dead load + imposed load + wind load	($\gamma_D = \gamma_L = \gamma_w = 1.2$).

In buildings of simple construction wind loading is assumed to be transferred from the steel frame into the horizontal bracing system as illustrated in Figure 10.3. Thus the designer will normally only need to consider case 1 for the frame – although other cases may govern the design of other parts of the structure, e.g. the holding-down bolts. Moreover, it will normally be the case that applying full load to the whole of the frame will give the most severe conditions. Exceptions may occur at internal columns, for which larger unbalanced moments (but smaller compressive loads) will result if the imposed load is omitted on some beams as shown in Figure 10.7.

At first sight the concept of moments being produced in columns as a result of vertical loading on beams may seem at variance with the

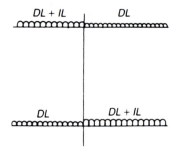

Figure 10.7   Loading arrangement to induce large moments at internal columns.

theoretical basis of simple construction illustrated in Figure 10.1(b). However, because real joints will not function as perfect pins – because they will possess some degree of rotational stiffness – the device already introduced in Section 2.1 (Figure 2.3) is used. Thus beam reactions are assumed to act at some distance from the face of the column – 100 mm or the centre of the bearing, whichever is the larger according to *Cl. 4.7.7* – with the result that a moment equal to the product of the reaction and this notional eccentricity must be designed for. It is because of this requirement that the concepts of unbalanced moment at internal columns may need to be considered. In the case of external and corner columns some moment about one or both axes must always be allowed for. It should, of course, be borne in mind that such members will normally attract a smaller share of direct vertical loading (see Figure 10.5).

Since joints are only required to transmit vertical shear, it follows that the only loading necessary for joint design will be the end reaction from the relevant beam. Reference to the examples of Tables 8.1 and 8.2 confirms this.

However, because of concern about progressive collapse in multistorey construction, i.e. ensuring that in the event of a local failure of one or more members in a particular region the integrity of the structure as a whole will be preserved, BS 5950: Part 1 does impose certain additional requirements on connections. These are basically to ensure that buildings are properly tied together in the horizontal plane [3]. Thus all buildings must meet the requirements of *Cl. 2.4.5.2*, which requires that the beams be capable of functioning as a two-way grid holding the columns in place. A design tie force of 75 kN (40 kN at roof level) is specified for each connection. In practice this is not onerous and can be achieved with two M20 bolts. The concept is thus one of catenary action, with the beams acting as ties, albeit at gross deflections if necessary.

In the case of buildings of more than five storeys the additional and potentially far more onerous requirements of *Cl. 2.4.5.3* must be satisfied. The magnitudes of the design tie forces must now be related to the beam reactions. In particular, with widely spaced and/or long span beams very large design tie forces are possible. It is, however, important to remember that for design purposes the end connections should be checked for tying forces acting separately from the normal loads, not in combination with them. Moreover, gross distortions are acceptable, so the problem is one of a truly ultimate condition, i.e. bolt heads pulling through holes in end plates, welds fracturing etc. A special procedure for checking the tensile resistance of beam to column connections for this particular limit state is therefore available [4].

10.1.4   Serviceability

Table 2.3 lists the three serviceability limit states considered by BS 5950: Part 1.

Some elementary material on corrosion was provided in Section 1.5; for the designer the main requirement is the selection of a corrosion protection system that is appropriate for the in-service conditions of the structure. Ordinarily this will reduce to the selection of a suitable paint system based on the guidance given in BS 5493. Up to date guidance is provided in a British Steel document [5], whilst Chapter 35 of the Steel Designers' Manual [6] explains the fundamental process in engineering terms.

Vibration due to wind loading is unlikely to be a problem for normal buildings since their natural frequencies will not be close to the excitation provided by wind gusts. This will not necessarily be the case for the other structural types, e.g. tower design is frequently largely controlled by considerations of wind action [7, 8]. One potential problem for buildings is vibration in long span floors, particularly when these are required to support sensitive computing equipment. For guidance on the best ways of avoiding such problems reference should be made to the recently published SCI Design Guide [9].

All structures deflect and thus an assurance that in-service deflections will not impair the function of a structure is as necessary as the provision of adequate strength under ultimate loads. This is traditionally provided by ensuring that the deflections calculated by basic linear elastic theory for a suitable representation of the structure do not exceed certain specified limits. For steel structures such calculations are normally made for a line model or 'wire frame' of the steel frame only. Moreover, it is usual to determine the serviceability deflections under imposed loads only (at $\gamma_L = 1.0$), since it is the variation in deflections in service that will be principally responsible for cracking in plaster ceilings etc.

For simple construction, since lateral loads are assumed to be carried by the bracing, this effectively reduces to calculating beam deflections. Since simple supports are usually assumed, the resulting values are likely to be somewhat larger than those observed in the real structure. This should be kept in mind when the size of the calculated deflections suggests a need for precambering of beams to reduce dead load deflections, i.e. providing each beam with a small initial upward deflection so that under the action of the dead load (principally the concrete slab) the beams are approximately level. Such an arrangement, in addition to being an additional operation adding about 10% to the cost of the steelwork, complicates detailing of connections. Expressions for the maximum deflection for some basic cases of simply supported beams are provided in Table 10.1; more comprehensive lists are available [6]. Because of the assumption of linear elastic behaviour, the principle of superposition may be used to combine load cases.

10.2   Continuous construction

The term 'continuous construction' was introduced in Section 2.1 to describe the class of steel frames in which the types of joint employed were able to maintain virtually unchanged the original angles between adjacent members. Since such joints are capable of transmitting substantial moments, the behaviour under load of these frames is more complex than that of the alternative 'simple construction'. In particular, member forces, i.e. moments, shears and thrusts, cannot now be obtained directly using simple statics. In principle, of course, utilizing continuity is structurally more efficient because of the greater participation of all parts of the structure in resisting the applied loads. Whether or not the resulting structure will actually be superior, i.e. more economic, more robust, etc., than the simply designed alternative is a complex question that was discussed briefly in Section 2.2.

For simple design the procedures described in Chapters 3–6 enable individual members to be selected; they may also be used for members in continuous structures providing the internal forces in these members have been calculated properly. This may be done on an elastic basis using any suitable method, such as moment distribution, slope deflection, and matrix stiffness [10], or providing certain restrictions are observed, using plastic theory [11, 12]. The second approach utilizes the ductility of steel, as discussed in Section 1.2, to permit redistribution of moments after the attainment of maximum capacity locally in the most highly stressed member. Since its use assumes the strength of the structure to be governed by a particular mode of collapse, it is necessary to eliminate the possibility of premature failure by other means, for example through the use of strict geometrical limits to control buckling.

Some indication of the types of joint suitable for use in continuous construction has already been provided in Chapter 8. Since these are normally more complex than those used for simple construction their use will involve more fabrication and they will therefore be more expensive. Erection on site may also be more difficult because of the tighter degree of fit-up required (more exact matching-together of the individual components). Thus continuous construction does have certain practical disadvantages which must be weighed against the more obvious structural advantages of using less steel and producing a generally stiffer, more robust structure.

Although special requirements may affect the choice for a particular structure, construction economics in the UK have tended to push the use of continuous construction into certain well-defined areas. Probably the most important of these is the use of portal frames of the type illustrated in Figure 10.8 for low-rise industrial buildings such as factory units and warehouses. It has been suggested [13] that these consume something approaching one half of the UK civil engineering market for structural

Table 10.1   Deflections of simply supported beams

Case	Maximum deflection

$$\frac{1}{48}\frac{WL^3}{EI}$$

$$\frac{1}{6}\frac{WL^3}{EI}\left[\frac{3a}{4L}-\left(\frac{a}{L}\right)^3\right]$$

$$\frac{1}{48}\frac{WL^3}{EI}\left[\frac{3a}{L}-4\left(\frac{a}{L}\right)^3\right]$$

this is value at the centre; it is always within $2\frac{1}{2}\%$ of the maximum value.

$$\frac{1}{8}\frac{ML^2}{EI}$$

$$\frac{5}{384}\frac{WL^3}{EI}$$

$$\frac{1}{96}\frac{Wa}{EI}\left[3L^2-2a^2\right]$$

steelwork. As a result, their design has become a highly refined and competitive process. Other important areas include the use of continuous beams in situations where limiting deflections or minimizing construction depth are important – multistorey frames for which considerations of access and flexibility in utilizing the internal space make the use of bracing unacceptable, and for highway bridges where continuity over the supports provides a better riding surface.

10.2.1   Elastic design of continuous structures

The design of continuous steel structures on an elastic basis consists essentially of applying the methods described in the preceding chapters using member forces calculated in a way that recognizes the effects of continuity. Because of the greater degree of interaction between different parts of the structure it may be more difficult to identify the exact load case corresponding to the most severe condition in each individual member. This is likely to be particularly true for members subject to combined loading for which design has to be based on some form of interaction approach, for example the support region of a continuous plate girder where coincident shear and moment values must be considered and beam columns carrying compression and unequal end moments. Fortunately BS 5950: Part 1 permits member forces to be obtained using linear elastic analysis, amplifying these where necessary to allow for instability effects. This has two important consequences for the designer: it enables him to sum the effects of different load cases using the principle of superposition and it gives him the opportunity to use standard frame analysis programs for extensive or irregularly shaped structures.

Figure 10.8   Typical portal frame structure during erection. (*Courtesy of Condor Midlands, Burton-on-Trent.*)

However, the Code gives little guidance on which arrangements of load are likely to be the most critical; *Section 5.1.2* merely refers to vertical loads 'arranged in the most unfavourable but realistic pattern for each element'. Horizontal loads need only be considered to act in conjunction with full vertical load. An earlier draft [14] did attempt to be more specific, including suggestions in the Commentary for use when considering multi-storey frames. Even the use of these does not guarantee that the most severe design condition for each member will be included [15]. When considering pattern loading it is not necessary to allow for variations in dead load as part of the pattern, i.e. γ_f should be taken as 1.0 for dead load throughout the structure.

Deflections under serviceability load conditions are generally subject to the same limits as for simply designed structures, as covered by *Cl. 2.5.2* and *Table 8*. An exception is made for portal frames and this is discussed in Section 11.4.

Provided an elastically designed continuous structure contains only members of compact cross-section (see Section 5.3.1), limited redistribution of moments is permitted. Thus within a beam, for example, the elastic moment diagram may be modified by up to 10% of the peak elastic moment, providing, of course, the resulting moments and shears remain in equilibrium with the applied loads. This concept may be thought of as a very limited recognition of the potential that exists within continuous

Light strusses, beams and columns for a warehouse.

structures to withstand loads in excess of those that require full member bending strength only at the most critical location. Since this is possible only if unloading does not follow the attainment of this local maximum strength, some limitation on cross-sectional geometry is required; this is the reason for limiting the process to compact sections.

10.2.2   Plastic design of continuous structures

The main differences between the behaviour of 'simple' and continuous steel structures can best be appreciated by considering a specific example.

Figure 10.9 shows the load versus central deflection relationship obtained from either a test or from a rigorous theoretical analysis for a simply supported beam. Three distinct phases may be identified:

1   OA elastic – linear relation between load and deflection;
2   AB elastic plastic – deflections increase at a progressively faster rate;
3   BC plastic – growth of large deflections at sensibly constant load.

Detailed consideration of the distribution of bending strain and bending stress at the most severely loaded cross-section at mid-span reveals that during phase 2 yielding is developing, while phase 3 corresponds to a state of full local plasticity (strictly speaking this is true only if certain simplifications are used which make it rather easier to appreciate the basic features of inelastic bending, for example the real stress-strain curve is approximated as a bilinear elastic – perfectly plastic relationship). The attainment of a fully plastic cross-section in bending is termed the formation of a 'plastic hinge'. The reason for the first part of the name is self-evident; the assumption of perfectly plastic material behaviour beyond yield means that the effective modulus for the material will be zero and thus the cross-

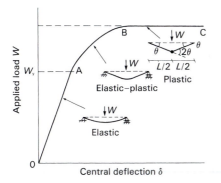

Figure 10.9   Behaviour of a simply supported beam. (BSC Teaching Project, Imperial College, 1985.)

section's effective value of *EI* will be zero. The beam will now behave rather as if a real hinge had been introduced at mid-span in that it will become a mechanism unable to resist any further increase in load – hence the horizontal load-deflection curve.

The behaviour of a continuous structure, for example the two-span continuous beam of Figure 10.10, will exhibit certain differences as indicated by the load versus mid-span deflection relationship shown as Figure 10.11. In this case the beam's response up to the formation of a plastic hinge at the point of maximum moment, based on the elastic moment diagram, will be basically similar to that of the simply supported beam. However, because of the redundancy in the system the appearance of this hinge at the central support transforms the beam not into a mechanism but into a statically determinate structure. Although it will function somewhat differently, in that the additional moments developed as a result of the application of increased load will be distributed differently, i.e. the moment at the central support cannot increase, it will nonetheless be capable of withstanding extra load.

Thus continuous structures possess the ability to redistribute load from the most severely stressed locations (plastic hinges) to less highly stressed areas. Only after sufficient plastic hinges have formed to convert the originally redundant structure into a progressively less redundant structure, then into a

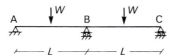

Figure 10.10   Two-span continuous beam.

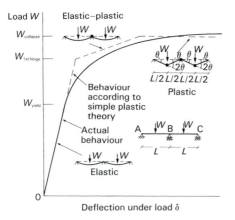

Figure 10.11   Load-deflection curve for a statically indeterminate steel beam. (BSC Teaching Project, Imperial College, 1985.)

statically determinate structure and finally into a mechanism, does collapse occur. Of course, utilization of this ability requires that premature failure by other means, such as elastic or plastic instability, or material breakdown due to insufficient ductility, be prevented. Therefore use of 'plastic design', as the exploitation of this redistributive phase after the formation of the first plastic hinge is termed, involves making certain assumptions:

1   The steel has adequate ductility as measured by the possession of a sufficiently long plastic plateau.
2   Plastic hinges, once formed, continue to rotate at a sensibly constant moment (taken as the full plastic moment of the cross-section).
3   Sufficient redistribution of moments can occur for the structure to fail by the formation of a plastic collapse mechanism.
4   Instability effects, either of the structure as a whole, of individual members or of the component plate elements of a member, do not prejudice the formation of the collapse mechanism.

Much of the material in BS 5950: Part 1 relating to plastic design is concerned with ensuring that these conditions are met. Readers who are unfamiliar with the methods of plastic analysis necessary for the determination of plastic collapse loads should consult a suitable textbook [9–11]; for portal frame structures, reference 16 provides an excellent commentary on all the key design issues discussed in detail in Chapter 11.

References

1   Brown, D.G. (1995) *Modelling of Steel Structures for Computer Analysis*, SCI Publication P. 148.
2   British Standards Institution (1996), *BS 6399: Parts 1, 2 and 3*, Loading for Buildings, London.
3   BCSA/SCI (1993), *Joints in Simple Construction, Vol. 1: Design Methods*, SCI, p. 207.
4   Advisory Desk (1990), *AD060 accidental damage and AD063 structural integrity tying*, Steel Construction Today, 4(2) 59 and 4(4) 128.
5   British Steel (1996), *The Prevention of Corrosion on Structural Steelwork*, British Steel.
6   Owens, G.W. and Knowles, P.R. (1992), *Steel Designers Manual*, 5th edn, Blackwell Scientific Publications, London.
7   British Standards Institution (1999) BS 8100, *Lattice Towers and Masts*, London.
8   CIRIA (1980) *Wind Engineering for the Eighties*, London.
9   Wyatt, T.A. (1989), *Design Guide on the Vibration of Floors*, SCI and CIRIA, Publication No 076.
10 Coates, R.C., Coutie, M.G. and Kong, F.K. (1987), *Structural Analysis*, 3rd edn, Van Nostrand Reinhold (UK), Wokingham.
11 Neal, B.G. (1977), *The Plastic Methods of Structural Analysis*, Chapman and Hall, London.

12 Horne, M.R. (1979), *Plastic Theory of Structures*, 2nd edition, Pergamon Press, Oxford.
13 Morris, L.J. (1981), *A Commentary on Portal Frame Design*', The Structural Engineer, 59A(19), 394–402.
14 British Standards Institution (1978), *B/20, Draft Standard Specification for the Structural Use of Steelwork in Building: Part 1: Simple Construction and Continuous Construction*, London.
15 Yau, F., Hart, D.A., Kirby, P.A. and Nethercot D.A. (1983) *Influence of loading Patterns on Column Design in Multi-Storey Rigid-Jointed Steel frames*, in L.J. Morris (ed.) Instability and Plastic Collapse of Structures, Granada, London, pp. 232–42.
16 Davies, J.M. and Brown, B.H. (1996) *Plastic Design to BS 5950*, Blackwell Science.

Chapter 11

Design aspects of continuous construction

The general principles of the design of continuous structures using either an elastic approach or a plastic approach have been set out in the previous chapter. *Section 5* of BS 5950: Part 1 also provides more detailed guidance on certain aspects of the design of the main forms of continuous construction:

continuous beams
portal frames (single storey)
multistorey frames

In each case either an elastic or a plastic approach is possible.

When using elastic design much of what has already been written in this book remains applicable. Plastic design, on the other hand, may be used only providing certain restrictions are observed. Thus before dealing with its application to any particular type of structure, it is necessary to be clear on the conditions under which its use is valid.

11.1 Requirements for the use of plastic design

The use of plastic design relies on the ability of steel to accept strains considerably in excess of yield, so that those regions of the structure in which plasticity develops (plastic hinges) can maintain their capacity to carry load. Thus ductile behaviour is required (a) from the steel so that yield may develop fully over member cross-sections and (b) from the members so that full redistribution of moments may occur. In order to achieve this, limits must be placed on the type of steel used as specified in *Cl. 5.2.3.3* and the proportions of the members employed. Other safeguards on ductile behaviour are that regions containing plastic hinges should be fabricated to a high standard as set out in *Cl. 5.2.3.4* of BS 5950: *Part 1* and that the structure be subject to predominantly static loading.

Plastic design is permissible for all grades of steel listed in BS 5950: Part 2. If other grades are to be used they must satisfy the requirements on ductility given in *Cl. 5.2.3.3*. Possession of an adequate plastic plateau is clearly necessary for the development of yield over a cross-section. It is perhaps not immediately clear why an ultimate tensile strength significantly in excess of the yield strength is also required; this ensures that strain hardening will take place. Although this is not normally used explicitly in plastic design, its existence is essential for the proper development of plastic hinge action [1, 2].

Members containing plastic hinges must satisfy the limitations on flange and web proportions for plastic or class 1 sections presented in Table 11.1. These are sufficiently more restrictive than those for compact sections that a number of standard UBs and UCs, especially in the higher grades of steel, will not meet them. Such sections can be used in plastically designed structures only in areas where plastic hinges will not be required to form.

Use of plastic design requires that beams be capable of attaining and maintaining their full plastic moment capacity (reduced where necessary to allow for the effects of coincident shear, see Section 5.3.1b), and that beam columns carry loads which correspond to their reduced plastic moment capacity, see Section 6.1, etc. Since the required member strengths are prescribed in advance, it is necessary to ensure that premature failure by member instability does not occur. Members containing the amounts of plastic material necessary for the formation of plastic hinges are particularly susceptible to failure by buckling due to the large reductions in stiffness (flexural, warping, etc.) produced by the presence of these yielded regions. Thus quite severe limits on member slenderness are required if the desired form of behaviour is to be achieved. Several methods exist by which the stability of members in plastically designed

Table 11.1   Cross-sectional limits necessary to prevent local buckling in members required to participate in plastic hinge action (based on *Table 11*)

Type of element	Method of manufacture	Limiting proportions for plastic design		
		$p_y = 275$ N/mm²	$p_y = 355$ N/mm²	$p_y = 460$ N/mm²
Outstand element of compression flange	Welded	b/T ≯8.0	≯7.0	≯6.2
	Rolled	b/T ≯9.0	≯7.9	≯7.0
Internal element of compression flange	Welded or Rolled	b/T ≯28	≯25	≯22
Web	Welded or Rolled	d/t $≯\dfrac{80}{1 + r_1}$	$≯\dfrac{70}{1 + r_1}$	$≯\dfrac{62}{1 + r_1}$

$r_1 = F_c/dt\, p_{yw}$, where F_c = total axial compressive force on the cross-section.

structures may be ensured; these often result from rather different approaches to the problem. Prior to the publication of BS 5950: *Part 1* the most popular method in the UK was that of the Constrado publication *Plastic Design* [3]; this is based on the work of Horne [4–6]. In *Cl. 5.3.3* the Code gives an expression for the maximum distance between points of restraint L_u as

$$L_u \not> \frac{38r_y}{\left[\dfrac{f_c}{130} + \left(\dfrac{p_y}{275} \right)^2 \left(\dfrac{x}{36} \right)^2 \right]^{\frac{1}{2}}} \tag{11.1}$$

in which f_c = compressive stress due to axial load (N/mm^2)
$\quad\quad p_y$ = design strength (N/mm^2)
$\quad\quad x$ = torsional index, see Chapter 5.

For a beam ($f_c = 0$) of S275 steel having the fairly high value of x of 36, equation (11.1) gives a limit of $38r_y$ which is in line with values specified in several overseas codes. Although *Cl. 5.3* defines restraint as 'lateral restraint', the requirement is really to prevent instability by bracing the member against both lateral deflections and twist. Bracing should always be provided in the immediate vicinity of a plastic hinge; if the actual hinge position cannot be braced then the restraint should act at a distance along the member of no more than half its depth.

   Clause 5.2.3.7 requires web stiffeners to be provided at plastic hinges if substantial applied loads act in that immediate region, for example for a main beam supporting a single cross-beam which is transferring floor load into that main beam. Loads are regarded as substantial if they exceed 10% of the web shear capacity, as explained in Section 5.3.1c. The immediate region is defined as within $D/2$ of the plastic hinge point; the stiffening must also be located no further than this from the hinge point. Stiffeners should be designed as load-carrying stiffeners, as explained in Section 5.3.1b, with the additional requirement that for flat plates b/t should not exceed 9.

11.2   Elastic design of continuous beams

Implementation of the various considerations outlined in the previous section of this chapter for the design of continuous beams on an elastic basis can most easily be appreciated by means of an illustrative example.

Example 11.1

Figure 11.1 shows a three-span continuous beam subject to a total design load of 75 kN/m over its whole length. Check whether a 457 × 191 UB 89 in S275 steel would be a suitable section. It may be assumed that the beam is adequately braced against lateral deflection and twist over its whole length.

Solution
Elastic analysis using, for example, the moment distribution method [7] gives the bending moment diagram of Figure 11.2. From this the maximum moment is 584 kN m. Since the beam is fully laterally restrained, obtain its bending strength directly from *Table 9* as $p_y = 275$ N/mm^2. The section under consideration is compact according to *Table 11*; therefore required section modulus is

$$S_x \nless 584 \times 10^3/275 \text{ cm}^3 = 2124 \text{ cm}^3$$

From section handbook for a 457 × 191 UB 89, $S_x = 2014$ cm^3 and this section is therefore inadequate. A stronger section is needed such as a 456 × 191 UB 98, for which $S_x = 2232$ cm^3.

Since the original section is close to being satisfactory (it is less than 5% understrength) and is compact, it is worth exploring the idea of limited redistribution of moments as permitted by *Cl. 5.2.2*. Figure 11.3 presents a modified bending moment distribution in which the support moments at B and C have been assumed to be reduced by 10% with corresponding increases in span moments.

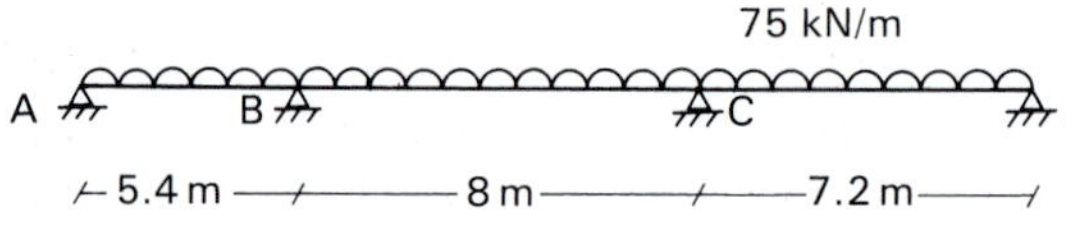

Figure 11.1 Continuous beam of Example 11.1.

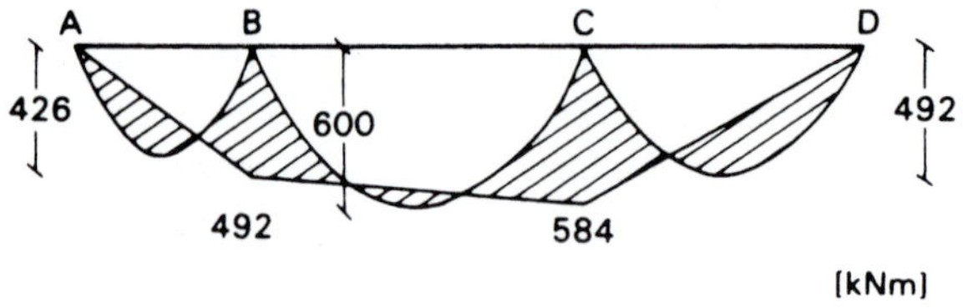

Figure 11.2 Elastic BMD for beam of Figure 11.1.

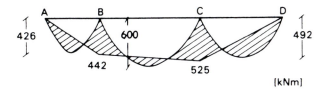

Figure 11.3   10% redistribution of elastic BMD for beam of Figure 11.1.

Solution
Maximum moment is now 525 kN m which requires

$$S_x \not< 525 \times 10^3/275 \text{ cm}^3 = 1909 \text{ cm}^3$$

Taking advantage of moment redistribution therefore permits the use of a
457 × 191 UB 89.

No allowance has been made in either set of calculations for possible
reductions in moment capacity due to the presence of high coexistent
shear forces. This is a topic to which greater attention must be paid when
designing continuous structures since the support regions will often be the
critical sections. It is left to the reader to verify (using the procedure of
Cl. 4.2.3 and *Cl. 4.2.5*) that the full value of S_x may be used in this example.
The assumption that the beam of Example 11.1 was fully laterally
restrained meant that design could be based on its full moment capacity. A
decision on the appropriateness, or otherwise, of this assumption is often
difficult for continuous beams. Indeed, because the patterns of moments
will be more complex than for simply supported beams, all considerations
of lateral stability will normally be less straightforward. Referring back to
Chapter 5, it will be recalled that lateral instability involved both lateral
deflection and twist and that an effective way of preventing its occurrence
was by bracing the beam's compression flange, for example the floor slabs
in a building can often be relied upon to laterally restrain simply supported floor beams. For continuous beams both flanges will normally be in
compression over part of the beam's length between supports. Figure 11.2
illustrates the common situation in which the bottom flange will be in
compression in the regions adjacent to the supports. A generally accepted
codified method of assessing the susceptibility of this region to lateral-
torsional instability is not presently available, although the topic has in the
context of composite beams received considerable study at the research
level [8, 9]. For hot-rolled sections it seems reasonable to assume that the
stiffness of the web will be sufficient to transfer the positional restraint
provided to the top flange by the slab to the whole beam. This may not be

the case for plate girders with more flexible webs. Doubtful cases should be checked using the U-frame approach of BS 5400: Part 3.

However, results from references [8] and [9] suggest this approach to be very conservative, indicating that for elastic design using rolled sections sufficient restraint is available from properly connected slabs for design to be based on the use of the section's full moment capacity at the support, i.e. lateral-torsional buckling will not be a design consideration.

In situations where even the top flange cannot be regarded as laterally supported such as continuous crane girders, and beams not positively attached to floor systems, the beam's moment capacity must be obtained using the procedures of *Cl. 4.3*. In such cases it will often be advantageous to take account of the less severe moment diagram by using the general expression for the m_{LT}-factor given in Table 18.

11.3   Plastic design of continuous beams

The application of plastic design to continuous beams involves two separate steps:

1   selection of a suitable section based on considerations of the plastic collapse mechanism;
2   ensuring that no other form of failure prevents the attainment of this collapse mechanism.

The first step requires the use of one of the standard techniques for plastic analysis [1–3, 11] such as the mechanism method or the reactant moment line method, while the second can largely be covered by compliance with the appropriate parts of *Section 5* of BS 5950: Part 1.

Plastic collapse may occur in either an internal or an end span of a continuous beam. It is therefore necessary to consider the two cases illustrated in Figure 11.4: a fixed-end beam and a propped cantilever. In both cases collapse will occur when sufficient plastic hinges have formed to turn the original statically indeterminate structure into a mechanism. Figures 11.4b illustrate these mechanisms while Figures 11.4c give the bending moment diagrams at collapse. The use of these results is illustrated by the following example.

Example 11.2

Repeat Example 11.1 using plastic design.

Solution
Referring to Figure 11.1, either span BC or span CD will govern the design. For BC, required $M_p = wL^2/16 = 75 \times 8^2/16 = 300$ kN m

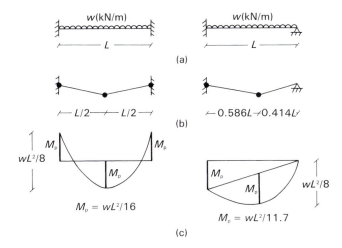

Figure 11.4   Plastic collapse of internal and end spans of continuous beams: (a) support arrangements; (b) plastic collapse mechanisms; (c) bending moments at collapse.

For CD, required $M_p = wL^2/11.7 = 75 \times 7.2^2/11.7 = 332$ kN m
Span CD governs, and taking $p_y = 275$ N/mm^2
Required $S_x = 332 \times 10^3/275 = 1207$ cm^3

Since S_x provided $= 2014$ cm^3, section is clearly adequate; based on considerations of moment capacity alone a 457×152 UB 60 for which $S_x = 1284$ cm^3 would be adequate. This is some 40% lighter.

The moment diagram at collapse is given as Figure 11.5. This shows that plastic hinges would occur at C and within span CD. Without resorting to complex elastic-plastic calculations it is not possible to define precisely the remainder of the reactant moment diagram and thus the exact span moments within AB and BC. However, the moment at B must lie between the value that would just permit a plastic hinge to form within BC and M_p (for which BC would remain elastic). In either case span BC would not collapse as a third plastic hinge would be necessary to produce a mechanism.

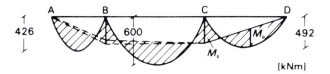

Figure 11.5   Bending moments at collapse for beam of Figure 11.1.

Reference to Table 11.1 shows that the revised suggestion of a $457 \times 152 \times UB\ 60$ is within the cross-sectional limits for plastic hinge action. To check whether the effect of shear will reduce its moment capacity refer to *Cl. 4.2.5.*

$$P_v = 0.6 \times p_y \times D \times t$$
$$= 0.6 \times 275 \times 454.7 \times 8.0 = 600 \text{ kN}$$

No reduction in M_p is necessary if the applied shear is less than 60% of this, i.e. max. $V \not> 360$ kN. Referring to Figure 11.5,

Shear to right of $C = 75 \times 7.2/2 + 332/7.2$
$$= 316 \text{ kN and no reduction in } M_p \text{ is required.}$$

11.4   Elastic design of portal frames

Figure 11.6 illustrates a variety of different portal frames of the type used as the main frames in single-storey buildings. In each case the connections between the columns and the inclined rafters must be capable of transmitting bending moments if the structure is to resist horizontal loading by frame action. If such structures are to be designed elastically it will therefore be necessary to conduct a suitable analysis (or series of analyses if multiple load cases are being considered). Although traditional methods may be used, the somewhat complex geometry of the sway deflections associated with joint translation means that programs based on the matrix stiffness method are often employed nowadays.

Once the individual member forces – moments, shears and axial loads –

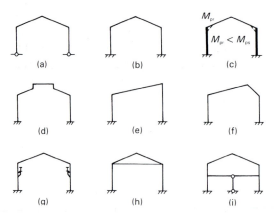

Figure 11.6   Types of single-storey portal frames: (a) pinned-base; (b) fixed base; (c) heavier column sections; (d) monitor roof; (e) lean-to; (f) north light; (g) including crane; (h) tied; (i) intermediate floor. (*After reference 11.*)

have been determined, selection of appropriate sections proceeds very much as for members in simply designed frames. Both the columns and the rafters will normally be subject to a combination of moment and compression; they should therefore be designed as beam columns using the procedures of Chapter 6.

When considering lateral-torsional stability, either *Cl. 5.3* or, as an alternative to the use of the methods of Chapter 6, the use of the more favourable method of *Appendix G* may be used. This makes some allowance for the restraining effects of the purlins, sheeting rails and cladding attached to the outer flange of the main frame members where this flange is the tension flange. For those parts of the frame where the outer flange is in compression, for example the rafters adjacent to the apex under vertical loading, it is, of course, only necessary to check stability between points of effective lateral restraint, i.e. between purlin points in most forms of construction.

A simpler alternative to the full procedure of *Appendix G* consists of using the provision of *Cl.5.3.4*. Providing certain restrictions on haunch proportions are observed, for the eaves region the limiting spacing for compression flange restraints is:

$$L_s \not> \frac{620 r_y,}{K_1 (72 - 100/x^2)^{\frac{1}{2}}} \text{ for S275 steel} \qquad (11.2)$$

If S355 steel is used the values of 620 and 72 must be increased to 645 and 94. The value of K_1 depends on haunch proportions, being unity if no haunch is provided and increasing to 1.4 for a haunch addition of double the depth of the rafter section.

The ability of portal frames to resist sway deflections, either due to the direct action of horizontal loads or unbalanced vertical loads or as a result of vertical loads exerting a destabilizing influence by acting through the out-of-plumb lateral deflections caused by lack of verticality of the columns, derives principally from frame action. Although methods exist [12] for taking account of the stiffening effects of the cladding, even if such a design approach is used, a certain basic level of overall stability of the bare frames is still required. For single-bay, single-storey portals designed to a sensible limit on serviceability deflections, sway instability is unlikely to be a problem. BS 5950: *Part 1* does not quantify this limit for elastically designed portal frames, other than to remind designers that a suitable value depends upon the form of cladding adopted.

Based on the results of a survey of the limits actually used by designers in Australia, Woolcock and Kitipornchai [13] have suggested a set of limits for lateral deflections for portals, some of which are given in Table 11.2. These distinguish between building construction and building use, items that are quite appropriate when considering service load behaviour.

Table 11.2   Recommended lateral deflection limits for portal frames. (After reference 13)

Building type	Recommended limit	Comments
Industrial, steel sheeting, no internal partitions against external walls or columns	$h/150$	
Industrial, steel sheeting, no internal partitions against external walls or columns, gantry crane	$h/250$	Take h to crane rail Use $h/300$ for heavy cranes
Industrial, external masonry walls supported by steelwork	$h/250$	
Agricultural	$h/100$	

A rather more fundamental treatment of the subject is provided in an advisory note [14], which was intended to supplement the 1990 version of the Code. This makes the important point that differential deflection, i.e. the movement of one frame relative to its neighbour, is at least as important a consideration as absolute deflection. A particular example of this, is the deflections of the first actual portal in from the usually very stiff gable end frame in the building.

In all matters of acceptable deflections under service conditions, the overriding consideration must be that the building function adequately under normal operating conditions. Thus situations involving the use of large doors, cranes, vibrating machinery etc., may require special consideration. Advice from owners and/or suppliers is normally relevant.

11.5   Plastic design of portal frames

Probably the most widespread application of plastic design is to single-storey portal frame structures. Often these will be the basic pitched-roof variety of Figure 11.6(a, b). Indeed the popularity of this form of construction has been sufficient to justify the publication of several specialist texts [10, 11, 14]. In the absence until publication of BS 5950; Part 1 in 1985 designers have placed considerable reliance on the approach of the Constrado publication on plastic design [3, 16]. This in turn makes use of earlier publications on the subject by the BCSA [4, 17]. Designers wishing to use the rules of *Section 5* of BS 5950: *Part 1* to design portal frames plastically may find it helpful to refer to these earlier publications for a full treatment of the subject since the Code provides guidance on only a number of specific items. Alternatively, the recent authoritative book by Davies and Brown [18] provides an in-depth coverage of all important aspects of the subject. The actual application of plastic design involves the same two basic steps that were given at the start of Section 11.3 on continuous beams, viz. adequate strength, avoidance of secondary failures.

For fixed-based, pitched-roof frames, collapse under vertical load normally occurs in the mode shown in Figure 11.7. Four plastic hinges are needed to transform the frame into a mechanism (it originally had three redundancies, for example horizontal and vertical force and bending moment at the apex). Possible locations are the peaks of the elastic moment diagram, i.e. the joints, and some point within each rafter. Since vertical load is transferred from the cladding to the main frames at the purlin points, the rafter hinge(s) form at whichever of these corresponds to the point of maximum sagging moment in the rafter. This is usually one or two purlin points away from the apex. However, failure to locate the exact point, although it will lead to a violation of the fundamental yield condition of plastic theory [2], normally results in only a marginal underestimate of the required M_p value. Bearing in mind the discrete nature of the available section range, i.e. their values of S_x do not constitute a continuous spectrum, this is unlikely to prove very significant.

Application of the semi-graphical approach [11, 18] therefore leads to the distribution of moments at collapse shown in Figure 11.8. Knowledge of this permits the selection of a suitable section to provide the desired margin of safety. (When using a limit-states approach in which different

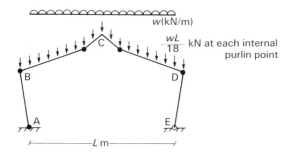

Figure 11.7   Typical plastic collapse mechanism for a fixed-base frame under vertical load.

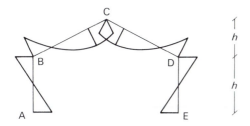

Figure 11.8   Bending moments at collapse for frame of Figure 11.7.

numerical values are required for the various components of the applied loading, such as $\gamma_f = 1.4$ for dead loads and $\gamma_f = 1.6$ for imposed load, it is necessary to work with the actual factored loads; earlier publications, which used a single global load factor – typically 1.7, often showed calculations arranged in such a way that the designer could select a section to give any desired value.)

Identification of the collapse mode associated with the lowest value of the collapse load is normally quite straightforward for single-bay frames. Of the three basic modes for fixed-base frames illustrated in Figure 11.9, mode 2 is possible only for very tall frames and/or large horizontal forces whilst mode 3 is likely only for high horizontal loading with negligible vertical load [3]. Similar behaviour is obtained for pin-base frames. Charts for the direct selection of a suitably factored value of M_p for either type of frame when subjected to a uniform vertical load plus a single horizontal eaves load are provided in reference [11]. These charts are useful for gaining an indication of the probable collapse mechanism as they permit the use of different load factors for the two types of loading. Doubtful cases can thus be identified and check fully using a more realistic representation of wind load [3, 18]. Because of the lower load factors permitted under combined loading (see Chapter 2), most designs will be governed by the gravity load case for which good estimates of the required section, assuming a uniform frame, may be obtained from

$$M_p = \gamma_L \frac{wL^2}{8} \left[\frac{1}{1 + h_2/h_1 + (1 + 2h_2/h_1)^{\frac{1}{2}}} \right] \text{ for pin-base frames (11.3a)}$$

$$M_p = \gamma_L \frac{wL^2}{8} \left[\frac{1}{1 + 0.5h_2/h_1 + (1 + h_2/h_1)^{\frac{1}{2}}} \right] \text{ for fixed-base frames}$$

$$(11.3b)$$

in which γ_L = global load factor
and h_1 and h_2 are defined in Figure 11.8.

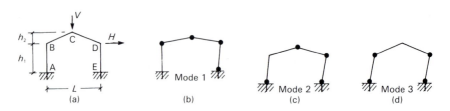

Figure 11.9 Basic plastic collapse mechanisms for fixed-base portal under combined load. (After reference 3.)

In deriving equations (11.3) [17] the rafter hinges have been assumed to form at the point of maximum moment for distributed loading, i.e. the discrete nature of purlin loading has been ignored; this affects the resulting value of M_p by only a very few per cent.

For single-bay frames having other geometries, for example those shown in Figure 11.6, for single-bay frames having other than the same section throughout, or for multistorey frames, the reader should consult references [1–3, 11, 18] for suitable methods of plastic analysis. These also explain the basis for the use of haunches – at either the eaves or the apex as shown in Figure 11.10 – as a means of improving frame strength and frame stiffness. Eaves haunches are often made by splitting the basic rafter section along a diagonal and are widely used in the UK, the additional fabrication costs being more than offset by both the reduction in material and the greater rafter depth available for constructing the eaves joint. Assuming that the eaves haunches shift the eaves plastic hinges to the column top immediately below the lower end of the haunch, as shown in Figure 11.11 for a fixed-based frame whose design is controlled by vertical load, the reduction in M_p is given by [3]

$$\frac{M_p \text{ with eaves haunch}}{M_p \text{ for uniform frame}} = \left(1 - \frac{a}{h_1}\right) \tag{11.4}$$

in which $a =$ depth of haunch (see Figure 11.10).

For typical geometries savings of between 5% and 10% on rafter size are possible. Because of the reduction in rafter moments the use of eaves haunches often leads to frames with lighter rafters than columns. Provision of a haunch at the apex does not affect the frame's basic strength (the collapse mechanism does not involve a plastic hinge at this point). However,

Figure 11.10   Use of haunches in portal frame construction. (*Reference 11.*)

Figure 11.11   Basic collapse mechanism (Mode 1) for haunched portal.

it does reduce overall frame deflections as well as providing greater depth for the apex connection.

The types of collapse mechanism which govern the design of most portal frames are such that virtually every member is required to participate in plastic hinge action. Therefore they should each meet the cross-sectional limitations for plastic hinge action of Table 11.1. Arguments based on the identification of the last hinge to form (at which no rotation is required which, at least in theory, suggests that merely satisfying the limits for a compact section would be appropriate) should be treated with suspicion due to the difficulty of being certain that this can actually be identified [19]. Factors such as settlement, variability of material strength between members, etc., while they do not necessarily affect the plastic collapse load significantly, can change the sequence of hinge formation.

Premature failure due to lateral-torsional instability may be avoided if torsional restraints are positioned according to equation (11.1). Purlins attached to the compression flange of a main member would normally be acceptable as providing full torsional restraint; where purlins are attached to the tension flange they should be capable of providing positional restraint to that flange but are unlikely (due to the rather light purlin/rafter connections normally employed) to be capable of preventing twist. Allowance for the limited benefit of tension flange restraint may be made by using the plastic version of the method of *Appendix G* of BS 5950: Part 1. This permits lateral restraint on such members to be spaced at a distance L_y given by

$$L_y \gg \frac{L_k}{\sqrt{m_t}} \left(\frac{M_{psc}}{M_{psc} + aF_c} \right)^{\frac{1}{2}} \tag{11.5}$$

in which a = distance of member axis to restraint axis (the effective position of the restraint axis may well lie beyond the member's tension flange at the line of the sheeting).

$$L_k = \frac{(5.4 + 600p_y/E)r_y x}{[5.4(p_y/E)x^2 - 1]^{\frac{1}{2}}}$$

Modifications to equation 11.5 are provided to allow for taper and/or non uniform moment. Prior to the publication of BS 5950: Part 1 the method given in reference [16], which is based on the original work of Horne [4–6], was widely used. Although not specifically mentioned by the new Code, it would appear to remain as an acceptable alternative.

Failure to meet whichever of the above criteria is selected can usually be rectified by a combination of knee braces to stabilize the main member's compression flange as shown in Figure 11.12 and a rearrangement of the

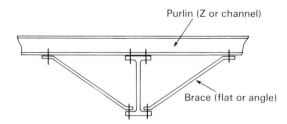

Figure 11.12   Knee-brace to compression flange of main member.

purlin spacing. The most critical area is normally the rafter adjacent to the eaves for which both modifications may be necessary. Although the methods for assessing lateral stability in BS 5950: Part 1 are permissible (according to that document) for tapered, i.e. haunched members, experimental evidence on the performance of haunches containing plastic regions [20] suggests that instability effects are particularly severe. Until a full solution to this problem is available it is probably advisable to follow earlier advice [15, 18, 19] and ensure that haunch regions remain elastic.

The foregoing discussion has assumed that the frame moments due to the applied loading may be assessed on the basis of first-order theory, i.e. a basic rigid plastic analysis is suitable for calculating the collapse load factor λ_p. For most sensibly proportioned frames this will be the case. However, recent trends towards the use of lighter rafters (due to the presence of substantial eaves haunches) coupled with flatter roof pitches – 6% slope is now common, sometimes reduced to as little as 2% [18], when 15% or 20% was once the norm [3, 19] – mean that it is no longer sufficient merely to assume that every arrangement may be treated in this way. The higher rafter axial loads resulting from the flatter pitch means that in-plane instability effects will sometimes reduce the actual in-plane failure load sufficiently below λ_p that a quantitative allowance must be made. *Cl. 5.5.4* presents the procedure for undertaking this. Three alternatives are provided:

a   Providing frame proportions meet the necessary restrictions, using the sway deflection check of *Cl. 5.5.4.2*
b   Amplifying the first-order moments based on the value of the elastic critical load for the frame using a Merchant-Rankine criterion to check whether this is necessary as described in *Cl. 5.5.4.4*
c   Conducting a full second-order, elastic-plastic analysis (only feasible if appropriate software available). This is the only method permitted for frames in which an eaves tie is used to reduce apex deflection; such an arrangement can very significantly increase the compressive loads in the rafters.

11.6   Elastic design of multistorey frames

When designing a multistorey frame in which continuous construction is to be employed it is first necessary to establish the means by which overall stability against sway effects is to be provided. Figure 11.13 illustrates the two basic alternatives – lateral bracing in the form of concrete shear walls, a central braced core or a series of braced bays or dependence on frame action. Of these the first alternative is generally claimed to be the more economic for the types of structures built in the UK. However, client's requirements for access and utilization of internal space is sometimes too restrictive for bracing to be used, in which case a sway frame is required.

Classification of multistorey frames as 'non-sway' or 'sway' is to some extent subjective as all structures deflect laterally under the action of horizontal forces. Indeed it is possible for lightly braced frames to deform more than laterally quite stiff, unbraced frames such as those with very heavy columns [21]. *Clause 2.4.2.6* of BS 5950: *Part 1* differentiates between the two classes by setting a limit on the horizontal deflection δ in any storey of a non-sway frame as

$$\delta \not> h/2000 \tag{11.6}$$

in which h = storey height.

Equation (11.6) should be used for clad frames for which the calculations have been performed on the bare frame. If the stiffening effect of the cladding has been included, or the frame will not be clad in its finished condition, then the limit should be halved. Frames which do not meet this condition must be designed as sway frames. When conducting this check it is not necessary to use actual loads; notional forces equal to 0.5% of the factored dead plus imposed vertical load applied horizontally are specified in the Code. This is because the principle of the check is to assess the susceptibility of the structure to overall frame instability [22] and not to try to assess its actual sway deflections.

For non-sway frames designed on an elastic basis member forces under both vertical and horizontal loading should be determined from a linear

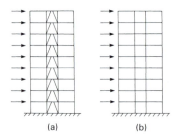

(a)   (b)

Figure 11.13   Non-sway and sway multistorey frames: (a) non-sway, (b) sway.

elastic analysis of the whole frame. This is most conveniently conducted using a standard frame analysis program. In the case of regular, rectangular structures it will normally be sufficient to isolate typical frames along and across the structure and to consider planar behaviour only. Structures with more complex plan geometries or those liable to be subject to significant unbalanced loading may require at least a limited three-dimensional analysis to assess correctly the importance of overall torsional effects. *Clause 5.6.2* of BS 5950: Part 1 permits two-dimensional analysis under vertical loading only to be undertaken using subframes of the type shown in Figure 11.14 as an alternative to analysis of the full frame. In conducting the analysis under vertical loading only, the Code reminds the designer of the need to consider 'the most unfavourable but realistic pattern for each element', without giving details of what these patterns should be. The earlier draft [9] suggested the arrangements of Figures 11.15 and 11.16 or, if subframe analysis was being used, those of Figures 11.17 and 11.18.

More recent research [23] suggests that the identification of the particular loading arrangement that leads to the most severe combination of member forces (moment, thrust and shear) in any individual member is more difficult, the patterns of reference [24] giving underestimates in several cases.

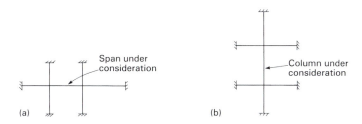

Figure 11.14   Subframes for use in multistorey frame design: (a) beam design sub-frame; (b) column design subfrace.

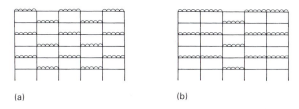

Figure 11.15   Load patterns for beam design: (a) maximum span moments (2 such patterns required); (b) maximum support moments (6 such patterns required).

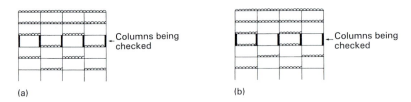

(a)

(b)

Figure 11.16   Load patterns for column design: (a) single curvature bending; (b) double curvature bending.

(a)

(b)

Figure 11.17   Subframes for beam design: (a) span moments; (b) support moments.

(a)

(b)

Figure 11.18   Subframes for column design: (a) single curvature; (b) double curvature.

While this may not be of great concern for residential buildings, for which imposed load will constitute only part of the total load and extreme variations of pattern loading are unlikely, it requires attention when dealing with certain storage or industrial buildings. Up to 10% redistribution of the elastic moments is permitted providing frame members are of compact cross-section and that redistribution does not lead to reductions in column minor-axis moments. Having determined member forces, suitable sections may be selected using the procedures already discussed in Chapters 4–6. In determining effective column lengths the restraining effects of the beams and columns immediately adjacent to the column being designed may be allowed for by using the limited frame shown in Figure 11.19 in conjunction with the effective length chart of Figure 11.20 [22, 25].

Sway frames should first be checked as non-sway frames using column effective lengths from Figure 11.20 under the most unfavourable combination(s) of vertical load. They should then be designed for the effects of sway by considering the full vertical load to act in conjunction with the horizontal loading, including the case of notional horizontal loading without wind. The second-order effects associated with sway deformations may approximately be allowed for by either:

1   using appropriately enhanced column effective lengths, see *Appendix E*;
2   amplifying the moments due to horizontal loading by a factor

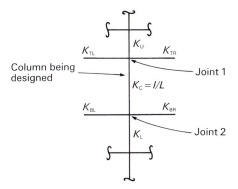

Figure 11.19 Restraint coefficients for a limited frame.

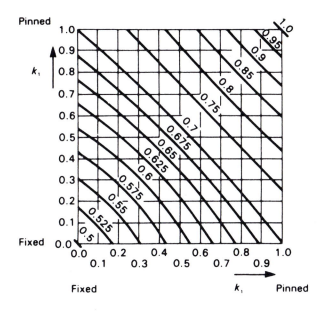

Figure 11.20 Effective length ratio for a column in a rigid jointed non-sway frame. (*After reference 22.*)

$$\lambda_{cr}/(\lambda_{cr} - 1) \tag{11.7}$$

in which λ_{cr} is the load factor for elastic instability of the frame.

When using method (2), column effective lengths equal to the storey height should be used. For the determination of λ_{cr} *Appendix F* outlines the approximate method of Horne [11, 22, 26] which used the sway deflections given by a linear elastic analysis of the frame under an appropriate set of horizontal loads to estimate the elastic critical load.

Metal decking waiting to form part of a composite floor.

11.7   Plastic design of multistorey frames

Provided the structure satisfies the requirement of equation (11.8) for treatment as a non-sway frame it may be designed plastically without specific consideration of its response to lateral loading. Determination of the plastic collapse load will require the application of one of the standard plastic analysis techniques [1–3, 10, 11]. Because the presence of significant amounts of plasticity in the columns can cause large reductions in stability unless the design is carefully controlled, it is normal practice to limit plastic hinge action to the beams. In such cases the application of plastic design effectively consists of treating the beams as continuous over the columns and designing for vertical load only. Two methods of doing this for an eight-storey frame example are presented in reference [3]. An alternative approach which makes use of continuity in both planes of the (assumed rectangular in plan) frame is the subject of a joint Institution of Structural Engineers and Institute of Welding report [27]. As a result of two series of full-scale tests [28, 29], improvements to the method of column design in that report have been proposed [30].

Plastic design of sway frames is a complex topic which is still the subject of much research. The central problem is the need to make suitable allowance for the effects of sway. Therefore before attempting to design on this basis it is necessary to acquire a proper understanding of the structural actions involved, for example by studying the relevant sections of references [2, 10, 11, 13, 21, 24]. Simply attempting to use the material of *Section 5.7.3* of BS 5950: Part 1 without this basis is likely to lead to misapplication of the rules.

The method of BS 5950: Part 1 uses the empirically based Merchant-Rankine formula [11, 22, 25] to assess the severity of sway effects. This formula is based on the premise that since collapse occurs by an interaction of plasticity and instability a good estimate of the failure load can be obtained from a knowledge of the simple plastic collapse load and the elastic critical load. The former gives the collapse load for very stocky frames (for which instability effects are negligible) while the latter provides a good indication of the strength of very slender frames (which effectively fail by elastic instability). Thus, providing λ_{cr} is greater than 10 for an analysis based on a bare frame that will subsequently be clad, or 20 for either an unclad frame or a clad frame in which allowance has been made in the analysis for the stiffening effect of the cladding, frame instability has negligible effect and design may be based on the simple plastic collapse load. Frames for which λ_{cr} is less than 4.6 in the first case or 5.75 in the second would experience such severe instability effects that their design must be based on a rigorous second-order elastic-plastic analysis [11, 30, 32]. For frames in the intermediate range the full analysis is not required providing the collapse load is taken as the simple plastic collapse load multiplied by a reduction factor which takes account of the limited influence of frame instability [22].

Exercises

1   Determine the maximum spacing between points of effective lateral restraint for a 305 × 102 UB 25 in S275 steel if such a section is used as a beam required to participate in plastic hinge action.

[0.60 m]

2   Check whether a 254 × 254 UC 89 in S355 steel carrying an axial load of 1400 kN over a free height of 2.5 m is capable of participating in plastic hinge action.

[No, L_m limit is 2.23 m]

3   Select a suitable UB in S275 steel to act as a 3-span continuous beam having spans of 7.2 m, 10.4 m and 6.5 m, assuming loading over the whole beam of 57 kN/m based on (i) elastic design, (ii) plastic design. In both cases you may assume full lateral restraint to all spans.

[457 × 191 UB 82 or 457 × 191 UB 74
if redistribution is used, 457 × 191 UB 67]

References

1 Neal, B.G. (1977) *The Plastic Method of Structural Analysis*, Chapman and Hall, London.
2 Horne, M.R. (1978) *Plastic Theory of Structures*, Pergamon Press, Oxford.
3 Morris, L.J. and Randall, A.L. (1979) *Plastic Design*, Constrado, London.
4 Horne, M.R. (1964) *The Plastic Design of Columns*, BCSA Publication No. 23.
5 Horne, M.R. (1956) The stanchion problem in frame structures designed according to ultimate carrying capacity, *Proc. Instn, Civil Eng.* **5**(1), Part III, 105–60.
6 Horne, M.R. (1964) Safe loads on I-section columns in structures designed by plastic theory, *Proc. Instn, Civil Eng.*, September, 137–50.
7 Coates, R.C., Coutie, M.C. and Kong, F.K. (1987) *Structural Analysis*, 3rd edn, Van Nostrand Reinhold (UK), Wokingham.
8 Johnson, R.P. and Bradford, M.A. (1983) Distortional lateral buckling of unstiffened composite bridge girders, in (L.J. Morris ed.) *Instability and Plastic Collapse of Steel Structures*, Granada, London, pp. 569–80.
9 Weston, G., Nethercot, D.A. and Crisfield M.A. (1991) Lateral buckling of continuous composite bridge girders, *The Structural Engineer*, **69**(5), 79–87.
10 Nethercot, D.A. and Lawson, R.M. (1992) *Lateral Stability of Steel Beams – Common Cases of Restraint*, Steel Construction Institute Publication No. 093.
11 Horne, M.R. and Morris, J.L. (1981) *Plastic Design of Low Rise Frames*, Granada, London.
12 Bryan, E.R. and Davies, J.M. (1982) *Manual of Stressed Skin Diaphragm Design*, Granada Publishing, London.
13 Woolcock, S. and Kitipornchai, S. (1987) Survey of deflection limits for portal frames in Australia, *J. Constructional Steel Research*, **7**(6), 399–418.
14 AD 090 (1991) *Deflection Limits for Pitched Roof Portal Frames*, Advisory Desk, Steel Construction Today, July, pp. 203–206.
15 Heyman, J. (1969) *Plastic Design of Structures*, Vols. 1 and 2, Cambridge University Press, Cambridge.
16 Morris, L.J. and Randal, A.L. (1979) *Plastic Design* (Supplement), Constrado, London.
17. Horne, M.R. and Chin, M.W. (1966) *Plastic Design of Portal Frames in Steel to BS 968*, BCSA Publication No. 29.
18 Davies, J.M. and Brown, B.H. (1996) *Plastic Design to BS 5950*, Blackwell Science.
19 Morris, L.J. (1981) A commentary on portal frame design. *The Structural Engineer* **59A**(12), 394–403 and **61A**(6, 7) 181–9, 212–21.

20 Morris, L.J. and Nakane, K. (1983) Experimental behaviour of haunched members, in (L.J. Morris ed.) *Instability and Plastic Collapse of Structures*, Granada, London, pp. 547–59.

21 Massonet, C. (1976) European recommendations for the plastic design of steel frames, *Acier-Stahl-Steel* (4), 146–53.

22 Kirby, P.A. and Nethercot, D.A. (1979) *Design for Structural Stability*, Granada, London.

23 Yau, F., Kirby, P.A. and Nethercot, D.A. (1983) Influence of loading patterns on column design in multi-storey rigid-jointed steel frames, in (L.J. Morris ed.) *Instability and Plastic Collapse of Structures*, Granada, London, pp. 232–42.

24 British Standards Institution (1978) B/20, *Draft Standard Specification of the Structural Use of Steelwork in Building: Part 1: Simple Construction and Continuous Construction*, London.

25 Wood, R.H. (1974) Effective lengths of columns in multi-storey buildings, *The Structural Engineer* **52**(7,8,9) 235–43, 295–302, 341–6.

26 Horne, M.R. (1975) An approximate method for calculating the elastic critical loads of multistorey plane frames, *The Structural Engineer*, **53**, 242–8.

27 The Institution of Structural Engineers and The Institute of Welding (1971) *2nd Joint Report on Fully Rigid Multi-storey Welded Steel Frames*, May.

28 Wood, R.H., Needham, F.H. and Smith, R.F. (1968) Test of a multistorey rigid steel frame, *The Structural Engineer* **46**(6), 107–19.

29 Smith, R.F. and Roberts, E.H. (1971) Test of a fully continuous multistorey frame of high yield steel, *The Structural Engineer*, **49**(10), 451–66.

30 Wood, R.H. (1973) *A New Approach to Column Design*, HMSO, London.

31 Vogel, U. (1983) Recent ECCS developments for simplified second-order elastic and elastic-plastic analysis of sway frames, *Third Int. Coll. Stability of Metal Structures*, Paris, 217–24.

32 Majid, K.I. and Anderson, D. (1968) The computer analysis of large multi-storey framed structures, *The Structural Engineer*, **46**, 357–69.

Chapter 12

Introduction to design for fire resistance

It is a legal requirement, stated formally in the *Building Regulations* [1], which govern all forms of building construction, that buildings in the UK be so designed as to exhibit an acceptable level of performance in the event of a fire. Essentially this is intended to ensure public safety rather than to safeguard the structure itself. Thus the main criteria are to prevent premature collapse, thereby permitting escape from the building, and to limit the spread of the fire, thus reducing the risk to surrounding properties and their occupants. The extent to which replacement of the actual steel frame would have to form part of the reinstatement of the building fabric in the event of a fire is very much a secondary issue.

Chapter 1 included some very general comments drawing attention to the fact that the basic material properties of steel used in structural design – its strength and stiffness – are adversely affected by increases in temperature beyond about 300 °C. Since significantly higher temperatures, perhaps approaching 1000 °C for the gas temperature but with rather lower steel temperatures in particularly severe cases [2], are possible in fires, it follows that proper consideration of the ways in which the integrity of the structure may be preserved are just as much a part of the structural designer's responsibility as is providing sufficient strength to resist the more traditional forms of loading such as floor loads, wind loads, etc.

At this stage it is as well to point out that there are several ways in which the necessary resistance of a steel frame structure to fire may be provided. By no means all of these call for protection of the actual steelwork – although this is often mistakenly seen as the only possibility. So-called 'active measures' include all types of monitoring and automatic extinction (of which the most common is a sprinkler arrangement designed to automatically release water throughout the structure when the outbreak of a fire is detected). Other forms of early warning of the existence of smoke and heat may be linked to direct summoning of the fire brigade and this may be supplemented by improved means of evacuation, incorporation of compartmentation so as to restrict fire spread, venting to release smoke and heat, etc. All of these measures reduce and may well even eliminate the need to actually protect the steelwork – the so-called 'passive approach'.

It is against this background that the Part 8 of BS 5950 [3] has been prepared and issued. In principle, this provides the engineer with the opportunity to employ a variety of methods for ensuring adequate fire resistance and gives detailed guidance on certain passive methods, i.e. assessing the inherent resistance of the steelwork and where necessary determining the levels of insulation needed.

12.1 Steel properties at elevated temperatures

The stress-strain behaviour of steel at elevated temperatures may be assessed in either of two ways:

1 using isothermal testing in which temperature is held constant and applied strain (stress) is increased;
2 using anisothermal testing in which applied stress is held constant and temperature is increased.

The second of these is generally regarded [4] as being the more representative of conditions in a building fire.

Figure 12.1 presents a set of such curves for S275A steels obtained from testing conducted by British Steel [5]. This type of information forms the

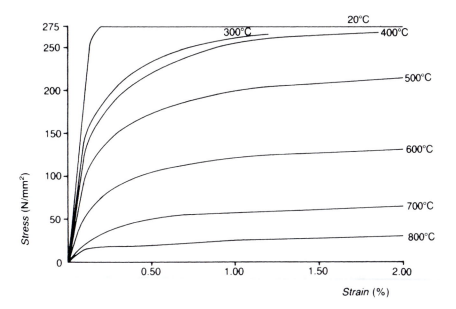

Figure 12.1 Elevated temperature stress-strain curves for S275A steel.

basis of the strength reduction factors in *Table 1*, the contents of which are applicable to all S275 and S355 material. *Table 1* may be used for tension, compression and shear. Although the tests of reference [5] revealed no significant effect of steel grade when the results are presented in the form of *Table 1*, this is no guarantee that steels of other compositions will behave similarly. Thus Part 8 requires testing if other types of steel are to be used. A more detailed discussion of the elevated temperature properties of steel is provided in the SCI Handbook to BS 5950: Part 8 [6].

Both Figure 12.1 and *Table 1* show that at temperatures in excess of about 550 °C steel loses some 50% of its room temperature strength. With working loads being of the order of 60% of the room temperature capacity ($\gamma_f \approx 1.5$), this suggests that under conditions of uniform heating, failure (in the form of excessive deflection) might be expected to occur at about this temperature. Thus the temperature of 550 °C is often referred to as the 'critical temperature', the implication being that once steel attains it, failure will follow more or less immediately. Whilst this is approximately correct for the particular case of uniform heating and full design load, most practical arrangements will not conform to both of these conditions, e.g. due to thermal shielding of part of the member as illustrated in Figure 12.2, with the result that failure will not occur when the maximum steel temperature reaches 550 °C. Indeed in particularly favourable situations

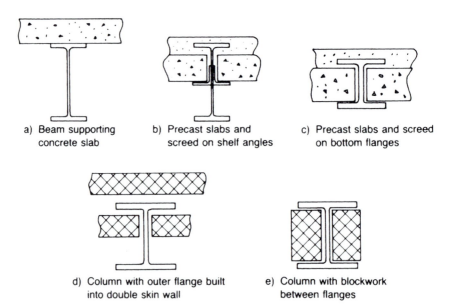

a) Beam supporting concrete slab

b) Precast slabs and screed on shelf angles

c) Precast slabs and screed on bottom flanges

d) Column with outer flange built into double skin wall

e) Column with blockwork between flanges

Figure 12.2   Forms of construction that inherently provide some degree of thermal shielding.

with steep temperature gradients and low load levels the hottest part of the steel may reach temperatures of the order of 800 °C without the member as a whole experiencing undue distress.

It is important to note from Figure 12.1 that the shape of the σ-ε curve alters as T is increased. In particular the sharp yield point characteristic of structural steels (see Figure 1.7) is replaced by a more rounded 'knee'. The stress at a suitable strain level, e.g. 0.2%, is then regarded as playing a similar role in characterizing material strength; it is often termed the '0.2% proof stress'. Similarly since the slope of the curves decreases with increasing T, material stiffness will also decrease, leading to greater deflections for the same load level. Both effects (strength and stiffness) will, of course contribute to the reduced performance of steel members at elevated temperatures.

12.2   Structural behaviour at elevated temperatures

Much of our understanding of the structural performance of steel members at elevated temperatures comes from the results of standard testing. In this the component is placed in a furnace, loaded by its working load, which should remain constant throughout the test, and subjected to controlled heating until failure occurs. Thus a 'standard fire' with the temperature-time relationship of Figure 12.3 as specified by BS 476 [7] and based on an international agreement [8] is used. As discussed in Section 12.3 this is not really representative of the temperature build-up in a real fire and is best regarded as a means of comparing performance rather than as an absolute measure. Methods of relating real fires to the standard

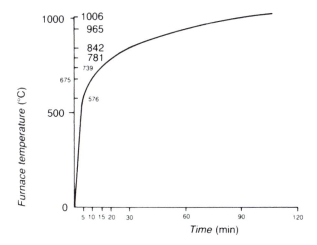

Figure 12.3   ISO 'Standard fire' temperature-time curve.

curve do exist [2] and since a large body of performance data to the standard fire are available, they still occupy a central role in the approach of Part 8.

12.2.1   Beams

Figure 12.4 illustrates the typical performance observed in a fire test on a steel beam. Deformation increases steadily but comparatively slowly for a considerable period of time (and thus temperature rise), AB. Quite suddenly deflections start to increase far more rapidly and a runaway deflection failure follows soon afterwards, BC. Traditionally a vertical deflection of $L/30$ has been taken as the failure criterion for beams, although $L/20$ or attaining a specified rate of increase of deflection are also used. For the behaviour shown in Figure 12.4 with a near vertical line for the latter part of the test clearly such distinctions would have very little effect.

If the beam supports a concrete slab so that it is partly shielded as indicated in Figure 12.2(a), then the improved performance of Figure 12.5 will be obtained. Further shielding, e.g. by supporting the floor slabs on shelf-angles as shown in Figure 12.2(b) or on the beam's lower flange as shown in Figure 12.2(c), produces additional gains, as indicated in Figure 12.5. Such behaviour leads to the concept of four-sided, three-sided, two-sided or one-sided attack [3], meaning that the steel is open to direct heating from all four, three, two or one side respectively. Clearly the more slowly the steel is able to heat up the greater the proportion of room temperature strength that is likely to be retained. An alternative view would be to regard the regions at the lower temperature as having retained a greater

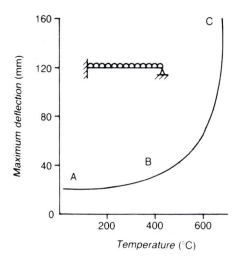

Figure 12.4   Typical performance in a fire test.

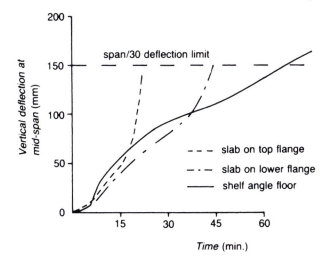

Figure 12.5   Behaviour in fire tests of beam types of Figure 12.2.

proportion of their room temperature strength; thus the effective cross-section for the determination of moment capacity will be of the form shown in Figure 12.6.

Appraisal of the range of available test data [9] has identified the section factor defined by

$$\text{section factor} = H_p/A \tag{12.1}$$

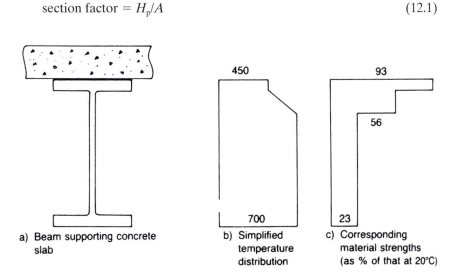

a) Beam supporting concrete slab
b) Simplified temperature distribution
c) Corresponding material strengths (as % of that at 20°C)

Figure 12.6   Determination of moment capacity.

in which H_p = heated perimeter in metres
(for an I-section subject to 4-sided attack $H_p \approx 4B + 2D$)
(for an I-section subject to 3-sided attack $H_p \approx 3B + 2D$)
A = gross cross-sectional area in m^2

as having a primary influence on the rate at which members heat up and thus lose strength. Low values indicate a squat type of section and a low rate of heating as illustrated in Figure 12.7. The influence of the H_p/A ratio as derived from tests on beams subject to three-sided attack, covering depths of between 203 mm and 838 mm, is shown in Figure 12.8. In this presentation the fire resistance time is defined as the time at which the beam deflection attained the limit of span/30.

One further factor found to have a significant effect on fire resistance time is the level of applied load on the beam. Not surprisingly beams carrying less than their full design load perform rather better than their fully loaded equivalents. Thus Figure 12.8 shows a direct relationship between

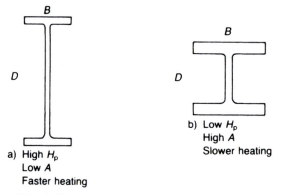

a) High H_p
   Low A
   Faster heating

b) Low H_p
   High A
   Slower heating

Figure 12.7 H_p/A ratio.

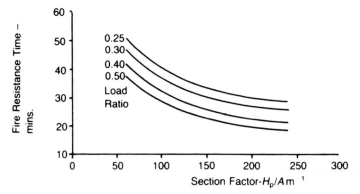

Figure 12.8   Effect of load and section size on fire resistance of unprotected beams fully exposed on three sides.

load ratio, defined as the ratio of the load actually carried to the value of the member's design capacity at room temperature, and fire resistance time. Clearly members in practice will have load ratios of less than unity, significantly so in those cases where basic strength is not the principal controlling factor in design.

12.2.2 Columns

Fire tests on columns, which for reasons of practicality in actually conducting the test tend to be restricted to comparatively stocky members, demonstrate qualitatively the same sorts of effects as described previously for beams. In this case there is less concern about deformations, the principal result simply being the time at which failure occurs. Alternatively one can think in terms of the reduced load-carrying capacity at any particular temperature. This leads to the sort of results of Figure 12.9 for uniformly heated bare steel columns [10].

Once again worthwhile improvements are possible if the column is thermally shielded over part of its cross-section, e.g. by having one flange built into a wall as shown in Figure 12.2(d). In this case the temperature gradient over the web of the section will induce deformations known as 'thermal bowing' that will be additive to both the initial lack of straightness and the lateral deflections produced by the load. A particularly simple example of a partially shielded form of construction consists of using concrete blocks to fill the space between the flanges as shown in Figure 12.2(e), thereby significantly reducing the rate of temperature build-up in the web of the steel section. This arrangement is covered in *Cl. 4.3.2* as well as in a separate design document [11].

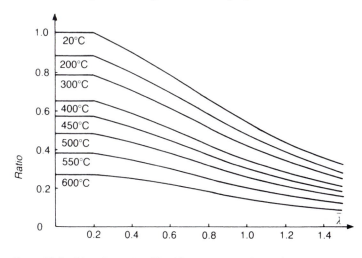

Figure 12.9 Non dimensional buckling curves at elevated temperature.

The importance of H_p/A ratio and load level is broadly the same for columns as it is for beams [4, 6]. Tables of H_p/A values for all UB, UC and SHS sizes for both three- and four-sided attack are provided in reference [6].

12.3   Fire engineering design

Design for the load case of fire is analogous to basic structural design to resist static loading such as dead and imposed loads due to gravity. Thus the necessary steps are:

1   assess the fire load;
2   assess the effect of this load on the steel frame;
3   determine whether the steel frame can safely resist this loading; if not take steps to ensure its integrity.

Figure 12.10 compares this process with the equivalent steps in conventional structural design.

12.3.1   Fire load

Assessment of fire loading may well amount to nothing more than selection of the requisite fire resistance period from the *Building Regulations* [1] for the particular class of structure. These list, in time periods of 30 minutes, 1 hour, 2 hours or 4 hours, the requirements in terms of building use, the main criterion for allocation being the need to ensure the integrity of the structure for long enough to provide for evacuation of the occupants. A secondary factor is the extent to which a severe fire may develop, e.g. fire tests in car parks [6, 12] have demonstrated quite dramatically that the combustible

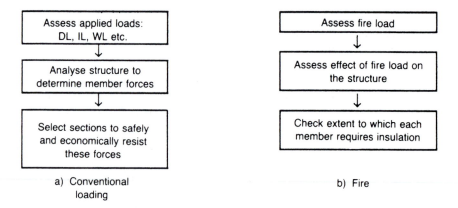

Figure 12.10   Main steps in design.

materials present (the cars) burn out surprisingly quickly so that after 30 minutes there is nothing left to fuel the fire and thus to generate heat.

A more sophisticated treatment consists of making a more general appraisal of the sort of fire that is likely to develop, the means of venting available for heat and smoke, available means of evacuation etc. [2]. Such an approach needs to be done on a case-by-case basis, if only because of the need to accurately assess the combustible material in the structure. It is of most use for buildings where a change of use is not really possible, e.g. hotels, schools, theatres, car parks, grandstands; it is less readily applied to a warehouse for which the stored material could be quite different were the building to pass to a new user. Such an approach treats the fire as a natural fire, i.e. it assesses the form of fire that would be most likely to actually occur in terms of the wood equivalent of the combustible material and a ventilation factor and then relates this to the BS 476 standard fire using the T-equivalent concept of Figure 12.11.

12.3.2   Effect on the steel frame

This step does not, at present, constitute an analysis in the same way that a structural analysis is usually undertaken to determine the forces generated in individual members as a direct result of the applied loads. Rather it requires consideration of the type of thermal shielding that may be present, the H_p/A factors for the members, the load ratios etc. – in short an assessment of those factors that may have some influence on fire resistance times. It is in this stage that the detailed guidance contained in Part 8 will be of most use.

If only 30 minutes fire resistance is required it may be possible that this can be supplied by the unprotected steelwork. *Table 4* gives the maximum H_p/A values for fully loaded members in the three categories:

1   beams supporting a concrete slab;
2   columns in simple construction;
3   columns with blocked in webs [11].

The effect of load ratio may be taken into account using the limiting temperature concept of *Cl. 4.4*. Thus *Table 5* lists for load ratios of between 0.2 and 0.7 the temperatures that may just be attained by various types of member corresponding to reaching their design condition, e.g. deflection of span/30 in the case of beams. *Tables 6 and 7* give the design temperatures for columns and beams respectively of varying dimensions that will be reached after 30 minutes, 1 hour, 2 hours or 4 hours. If the appropriate value from *Tables 6 and 7* is less than the value from *Table 5*, i.e. design temperature attained is less than limiting temperature at design condition, then the element may be assumed to be adequate.

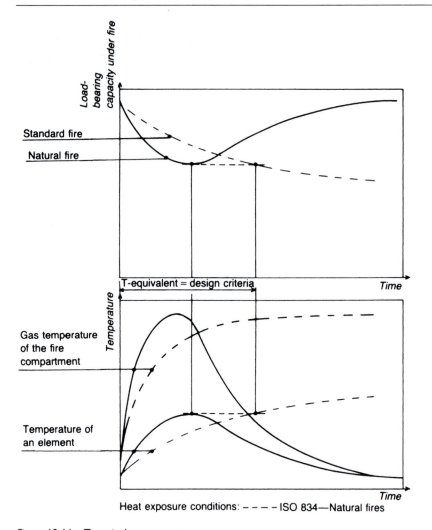

Heat exposure conditions: – – – – ISO 834—Natural fires

Figure 12.11   T-equivalent concept.

Example 12.1

A 533 × 210 × 122 UB supports a concrete floor. If 30 minutes fire resistance is required what fraction of its capacity can the beam safely carry? Assume live load is three times the dead load.

Solution

$$\gamma_{fL} = 1.6$$
$$\gamma_{fd} = 1.4$$

Combined load factor (room temperature) $= \frac{1}{4} \times 1.4 + \frac{3}{4} \times 1.6$
$$= 1.55$$

Portal frame building after fire damage

From *Table 2* use $\gamma_f = 1.0$ for DL and IL
∴ if fully loaded, load ratio $= 1/1.55 = 0.645$
From *Steelwork Design* Volume 1, flange thickness $T = 21.3$ mm
From *Table 7* for $T = 20.2$ mm design temperature $= 719\,°C$
From *Table 5* load ratio corresponding to limiting temperature of $719\,°C = 0.31$
∴ proportion of full capacity available $= 0.31/0.645$
$$= \underline{0.481}$$

In cases where this condition cannot be met, including those for which the acceptable load ratio would be too low, it is necessary to provide insulation to the steel so as to reduce the rate of heating.

12.3.3   Provision of insulation

Fire protection to the steelwork may be provided in a variety of ways [13]:

1   encasement in concrete or bricks;
2   spray protection;

3   board systems;
4   intumescent paints.

Concrete encasement is the longest established method of fire protection. Indeed it was the idea of using the concrete provided as fire protection in a structural sense that precipitated the concept of the composite column described briefly in Chapter 9. Compared with the newer methods, it adds significant load to the structure and because of the nature of the on-site concreting operation has an adverse effect on scheduling of the work. This may be improved upon if off-site pre-encasement is used but the trend nowadays appears to be towards the other lightweight approaches.

Sprays using asbestos-free materials, e.g. rock fibre or exfoliated vermiculite, may readily be applied to any shape. Since they are applied wet, the operation is messy and the results are such that the final appearance is not normally considered suitable for exposure in the building. This will be no problem if, for example, the results are to be concealed behind a suspended ceiling.

Boards provide a relatively clean, dry solution but the cost is rather greater than for sprays. They may well be the preferred solution, however, both because the surface is more acceptable and the fixing operation is less intrusive.

Intumescent paints are, as the name implies, painted on to the steel as a thin coating of almost 1 mm. When heated this releases a gas that inflates this layer into a thick carbonaceous foam that acts to insulate the steel. Intumescents are available in different forms, the more expensive can provide a 2-hour resistance, whilst the cheaper types are rated up to 1 hour and should not be used in damp environments such as swimming pools. More detailed accounts of their composition, mode of functioning and method of application may be found in the relevant manufacturer's literature. Guidance on all aspects of the use of intumescent coatings is available [14].

12.3.4   Calculation of insulation thickness

With the exception of intumescents, *Appendix D* provides the means to determine the necessary thickness t of insulation to provide a required period of fire resistance in the form of the equation:

$$t = k_i \left[\frac{H_p}{A} \frac{(I_t F_w)}{10^6} \right] \tag{12.2}$$

in which t is in m
k_i depends on the thermal properties of the insulation material
          (W/m per °C) and must be derived from tests in accordance with
          BS 476

H_p/A = section factor, see equation (12.1)

I_f = fire protection material insulation factor (m³/kW) obtained from *Table 16*

F_w = fire protection material density factor obtained from *Table 17*

Example 12.2

Repeat Example 12.1 to determine what thickness of insulation will be required if the beam has been designed to carry 85% of its moment capacity. If the fire rating is to be increased to 2 hours, by how much must the insulation thickness be increased?

Assume spray protection of density $\rho_i = 400$ kg/m³ and moisture content $c = 5\%$, take $k_i = 0.17$ W/m per °C and density of steel $\rho_s = 7850$ kg/m³.

Solution

At fire limit state load ratio = 0.85 × 0.645

= 0.55

From *Table 5* limiting temperature = 635 °C

Use equation (12.2) and *Appendix D*

From reference [6] H_p/A for spray on 3 sides = 110

From *Table 16* for a limiting temperature of 635 °C and 30 minutes fire resistance $I_f = 220$ m³/kW

$$\mu = \frac{k_i \rho_i (1 + 0.03c)}{\rho_s} \times \frac{I_f}{10^6} \times \left(\frac{H_p}{A}\right)^2$$

$$= \frac{0.17 \times 400 \, (1 + 0.03 \times 0.05)}{7850} \times \frac{220}{10^6} \times (110)^2 = 0.023$$

From *Table 17* $F_w = 0.975$

$$\therefore t = 0.17 \left[110 \times \frac{220 \times 0.975}{10^6} \right] = 0.0040 \text{ m}$$

$$= \underline{4 \text{ mm}}$$

From *Table 16* for a limiting temperature of 635 °C and 2 hours fire resistance $I_f = 1320$ m³/kW

$$\therefore t = 0.17 \left[110 \times \frac{1320 \times 0.975}{10^6} \right] = 0.0241 \text{ m}$$

$$= \underline{24 \text{ mm}}$$

12.3.5   Moment capacity method

As an alternative to the use of the limiting temperature approach of Section 12.3.2, beams whose temperature profile can be defined may have their fire resistance assessed on the basis of their available moment capacity corresponding to this particular temperature profile. In particular, *Appendix E* illustrates the application of this approach in detail for shelf-angle construction. A fully worked example for this form of construction is provided in reference [6].

Example 12.3

Determine the elevated temperature moment capacity of the beam of Example 12.1 assuming the temperature and strength distributions of Figure 12.12 with the change of strength in the web located 100 mm below the bottom surface of the upper flange.

Solution
The cross-section and strength distribution for the calculations is shown in Figure 12.12.

Longitudinal force capacity = $[(211.9 \times 21.3)\, 0.93 + (12.8 \times 100)\, 0.56$
$+ (12.8 \times 402)\, 0.56$
$+ (211.9 \times 21.3)\, 0.23]p_y$
$= [4198 + 717 + 2882 + 1038]p_y$
$= 8835 p_y \text{N}$

Position of neutral axis = $(4417.5 - 4198)/(12.8 \times 0.56)$
$= 30.6$ mm down from underside of top flange

$\therefore M_c$ $= [4198 \times (30.6 + 10.65) + 219.5 \times 15.3$
$+ 497 \times 34.7 + 1183 \times 270.4$
$+ 1038 \times 482.0] \times 275$
$= 279$ kN m

For this section S_x $= 3200$ cm^3
M_c (at 20 °C) $= 275 \times 3200 \times 10^{-3} = 880$ kN m
$\therefore M_c/(M_c)_{20\,°C}$ $= 279/880 = \underline{0.32}$

12.3.6   Partial encasement

In certain cases the form of construction will itself provide some degree of thermal shielding to the steel member, leading to enhanced performance in a fire. Several methods for allowing for specific arrangements that have been subject to scientific study are available.

For H-section columns, for which the spaces between the flanges have been filled with concrete blocks as illustrated in Figure 12.2e, BRE Digest

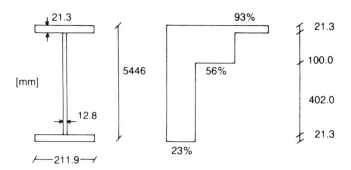

Figure 12.12   Cross-section and strength distribution for Example 12.3.

No. 317 [11] explains how 30 minutes fire resistance is normally achievable.

Sixty minutes resistance may be achieved if normal weight, poured concrete is fixed between the flanges by shear connectors attached to the web. Structural design does not regard the arrangement as a composite column under normal circumstances, but recognizes the thermal shielding in the event of a fire [15].

By supporting the floor slabs on angles attached to the beam web, as illustrated in Figure 12.2b, part of the steel section will be shielded by the concrete. Resistance periods of 60 minutes may readily be achieved [16].

Recently, considerable efforts have been made to develop the sort of concept shown in Figure 12.2c, in which rather shallow beams are used in an arrangement where the floor slab is attached to the lower flanges. Whilst this, clearly, does not provide the most efficient structural arrangement as a composite beam in the cold condition, the high degree of thermal shielding means that substantial fire resistance periods may readily be achieved. Moreover, since overall construction depth for the floor is reduced, cost savings on the complete structure are possible. Various schemes are available and Figure 12.13 illustrates two that have been developed by British Steel [17–19] for the UK market.

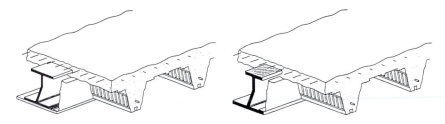

Figure 12.13   Fire restraint forms of floor construction.

For tubular sections used as columns, filling with concrete – possibly acting in conjunction with some bar or fibre reinforcement – enables resistance periods of up to 2 hours to be achieved. Such systems act in a combined way under load, the poorly conducting concrete heating less slowly, whilst the steel, even though it loses strength as its temperature increases, is still able to both confine the concrete and to prevent it from spalling. This form of construction is covered in the Part 8, as well as in specialist design guides [20, 21].

Metal deck composite floors of the normal types described in Chapter 9 inherently provide worthwhile levels of fire resistance without the addition of fire protection to the steel beams. Two methods for recognizing this when designing for the fire condition are available [22]. The first simply ensures that sufficient reinforcement is provided to the slab to ensure the required resistance period. Alternatively, allowance may be made for the redistribution of moments and by calculating plastic moment capacities at elevated temperatures, a fire engineering approach may be used to justify fire resistances of up to 4 hours.

12.4   Portal frames

Clause 4.5 covers the basic requirement for fire resistant design of portal frames. Because these are usually single-storey structures (neglecting mezzanine floors etc.), evacuation is much less of a problem and the main design consideration is to prevent damage to adjacent structures. Thus design requirements are sensitive to building location with respect to surrounding property, access roads etc. Rather than provide direct fire protection, it is usual to design for rafter collapse. *Clause 4.5* therefore requires that column bases and foundations be capable of resisting the forces and moments generated by a collapsing portal rafter. This aspect of design thus becomes a consideration of basic statics under the particular set of forces generated by the fire.

The various stages of behaviour leading up to collapse of a portal frame rafter are as follows [23].

1   The rafter expands due to temperature rise, producing small outward deflections of the eaves and upward deflections of the apex.
2   Fire hinges – similar in concept to plastic hinges but with a far lower moment capacity due to the reduction in material strength at elevated temperatures – form at the ends of the eaves haunches and at the apex.
3   Under the remaining vertical loading (assuming that a proportion of this is lost as a result of the fire) the rafter tends to form a 2- or 3-pinned arch. The axial thrusts developed as a result of this action induce column base moments in the opposite sense to those of stage 1.

4   The rafter falls below the eaves, may twist as a result of loss of lateral restraint from the purlins and acts as a catenary pulling inwards on the tops of the columns. This must be resisted by the base moments, the columns acting as vertical cantilevers.

For the more usual types of portal frame, i.e. constructed from hot-rolled sections, symmetrical etc., *Appendix F* provides formulae for these fire-induced base forces. In many cases it will be found sufficient to provide four holding-down bolts in the base, positioned outside the column profile. A more detailed account of this subject, including fully worked examples, is provided in an SCI publication [23].

References

1   *The building Regulations*
    (a) *Building Standards (Scotland) Regulations* (1990): Technical Standards, HMSO.
    (b) *Northern Ireland Building Regulations* (1994): Technical Booklet E, HMSO.
    (c) *Approved Document B* (1992) Building Regulations, HMSO.
2   Steel Promotion Committee of Eurofer (1990) *Steel and Fire Safety: A Global Approach*, Eurofer, Brussels.
3   British Standards Institution (1990) BS 5950: Part 8, *The Structural Use of Steelwork in Building: Code of Practice for Fire Resistant Design.*
4   Robinson, J.T. (1988) Fire Protection, in P.J. Dowling, R. Knowles and G.W. Owens (eds) *Structural Steel Design*, Butterworths, London, pp. 125–31.
5   Kirby, B.R. and Preston, R.R. (1988) High temperature properties of hot rolled structural steels for use in fire engineering studies, *Fire Safety Journal*, **13**, 27–37.
6   Lawson, R.M., and Newman, G.M. (1990) *Fire Resistant Design of Steel Structures – A Handbook to BS 5950: Part 8*, The Steel Construction Institute, Publication No. 080.
7   British Standards Institution (1987) BS 476, *Fire Tests on Building Materials and Structures*, Part 20: *Method of Determination of the Fire Resistance of Elements of Construction (General Principles).*
8   International Standards Organisation (1985) ISO 834, *Fire Resistance Tests – Elements of Building Construction.*
9   Robinson, J.T. (1989) Fire-resistant design of steel beams – recent developments in the UK, *Steel 2001*, 532–43.
10 Vandamme, M. and Janss, J. (1981) Buckling of axially loaded steel columns in fire conditions, *IABSE Proceedings*, P-43/81, pp. 81–95.
11 Building Research Establishment (1986) *Fire Resistant Steel Structures: Free Standing Blockwork – Filled Columns and Stanchions*, BRE Digest 317, December.
12 Bennets, I.D., Proe, D.J., Lewins, R. and Thomas, I.R. (1985) *Open Deck Car Park Fire Tests*, BHP Melbourne Research Labs.
13 British Steel (1998) *Fire Resistance of Steel Framed Buildings*, 1998 Edition, British Steel.
14 Steel Construction Institute (1996) *Structural Fire Design: Off-Site* Applied Thin Film Intumescent Coatings, Report RT 433.

15 Steel Construction Institute (1992), *The Fire Resistance of Web Infilled Steel Columns*.

16 Steel Construction Institute, (1993) *The Fire Resistance of Shelf Angle Floor Beams to BS 5950 Part 8*.

17 British Steel, (1998) *Design in Steel 2: Construction*.

18 British Steel, (1993) *Design in Steel 3: Fast Track Slim Floor*.

19 British Steel (1997) *SLIMDEK Construction*

20 British Steel (1996) *SHS Design Manual for Concrete Filled Columns: Part 2: First Resistant Design*.

21 Twilt, L. et al, (1994) *Design Guide for Structural Hollow Section Columns Exposed to Fire*, CIDECT.

22 Steel Construction Institute, (1998) *The Fire Resistance of Composite Floors with Steel Decking*, 2nd edition.

23 Newman, G.M. (1990) *Fire and Steel Construction: The Behaviour of Steel Portal Frames in Boundary Conditions*, 2nd edn, The Steel Construction Institute, Publication No. 087.

Index